BIOCHEMISTRY RESEARCH TRENDS

ZINC

CHARACTERISTICS, USES AND BENEFITS

BIOCHEMISTRY RESEARCH TRENDS

Additional books in this series can be found on Nova's website under the Series tab.

Additional E-books in this series can be found on Nova's website under the E-books tab.

EARTH SCIENCES IN THE 21ST CENTURY

Additional books in this series can be found on Nova's website under the Series tab.

Additional E-books in this series can be found on Nova's website under the E-books tab.

BIOCHEMISTRY RESEARCH TRENDS

ZINC

CHARACTERISTICS, USES AND BENEFITS

MASAYOSHI YAMAGUCHI
EDITOR

Nova Science Publishers, Inc.
New York

For permission to use material from this book please contact us:
Telephone 631-231-7269; Fax 631-231-8175
Web Site: http://www.novapublishers.com

LIBRARY OF CONGRESS CATALOGING-IN-PUBLICATION DATA

Library of Congress Control Number: 2011946006
ISBN: 978-1-61942-523-1

Published by Nova Science Publishers, Inc. † New York

CONTENTS

PREFACE

Zinc is one of the most abundant transition metals in the human body, and it has various benefits as an essential trace element in the regulation of the living body. Nutritional deficiency of zinc is known to induce many chronic diseases such as immune system disorders, rheumatoid arthritis, diabetes, obesity, atherosclerosis, cancers, bone growth retardation, osteoporosis and others. This book introduces the recent topics concerning the characteristics, uses and benefits of zinc. This book will be helpful in research development and also in the understanding of the various physiological-pathophysiological roles of zinc.

Chapter 1 - A study of the effects of Inhibition Characteristic of Vegetable Extracts on the corrosion susceptibility of locally processed Al-base alloy in some media concentrations with varying volume of vegetable extracts (*Telferia occidentalis* and *Piper guinenses*) have been carried out. Virgin aluminum alloys containing 1.0% and 2.0% zinc were cast and machined to cylindrical coupons of initial surface area averaging $11.64cm^2$. The pre-weighed samples were carefully immersed into 0.5 M and 1.0 M of the acid containing 50 cm^3 and 100 cm^3 of the extracts, respectively. The set-ups were allowed to stand for 28 days with a set withdrawn weekly for corrosion rate characterization. The trend of the results obtained were in tandem with that of passivating metals (or alloys) in simulated environments of initial steep rise in corrosion rate accompanied by a gradual decline as the exposure time increased. The vegetable extracts showed inhibition potentials with pronounced effects at higher extract concentrations. The best inhibiting property is found in Telferia occidentalis then, Piper guinenses.

Chapter 2 - A feasibility of ZnO-based materials was evaluated for a next-generation anode of Li-ion batteries. Composite thick-film electrodes of ZnO and Si were prepared by a gas-deposition method using an Si powder and a radial-shaped ZnO powder synthesized by a precipitation method. We evaluated anode performance of the composite electrodes, comparing with electrochemical properties of a ZnO electrode without Si. After the 20th cycle, the ZnO electrode exhibited a stable cyclability and a moderate discharge capacity of 160–220 mA h g^{-1}. The ZnO/Si composite electrode exhibited a better cycle stability after the 20th cycle and relatively-large discharge capacities exceeding the theoretical capacity of graphite in practical use until the 370th cycle. In contrast, the composite electrode containing of crushed ZnO showed flatter surface morphology and lower electrode performance in comparison with that using the radial-shaped ZnO. These results revealed that microstructure of the radial-shaped ZnO is very favorable for improving efficiencies of electrode reactions due to its high surface area, and is effective in accommodating the strains induced by Li-insertion and Li-extraction in/from Si.

Chapter 3 - Zinc is a vital element in the nutrition of human beings, animals and plants. Although most experimental data have demonstrated the adverse effects of Zn deficiency on human and animal health, there is still evidence to support the hypothesis that Zn toxicity can also result in deleterious health effects. Conflicting views have been expressed regarding the role of Zn in carcinogenesis. Some studies emphasize the fact that Zn deficiency causes cancers, while others highlight that Zn is involved in tumour growth and development of neoplastic transformation. These conflicting observations suggest that the role of Zn in different organs may be different. Based on the results of a micro-beam synchrotron X-ray fluorescence technique (micro-SRIXE) applied to the quantitative analysis of Zn, we correlated Zn concentration levels with the clinical stage of prostate cancer at the time of operation (described by the Gleason grade system). Serial sections of prostate tissues were collected from patients undergoing radical prostatectomy. Our results seem to be valuable in light of determination of the potential changes in Zn concentration as a diagnostic marker and its etiological involvement in the different stages of prostate diseases.

Chapter 4 - Atherosclerosis is a chronic inflammatory process, triggered by the presence of lipids in the vascular wall. The disease is associated with complex interactions among inflammatory cells, vascular elements, and lipoproteins, along with the expression of selected adhesion molecules and inflammatory cytokines. The strategy for protection against atherosclerosis has been a subject of intense research over the past decades. The important role of zinc has been recognized in various biochemical and physiological functions.

Early evidence indicates that zinc deficiency can cause growth retardation, delayed sex maturation, depressed immune response, and abnormal cognitive functions. Nutritional zinc deficiency is common in the developing countries. 30-40% of the elderly populations are estimated to have mild to modest zinc deficiency in the USA. Conditioned zinc deficiency is also present in many chronic diseases such as rheumatoid arthritis, diabetes, and cancers. The evidence from experimental, clinical, and epidemiological studies suggests that zinc has an atheroprotective function, potentially due to its anti-inflammatory and anti-oxidative stress effects, and other functions. However, the exact mechanism of zinc action in atherosclerosis is not fully understood. A large number of studies demonstrate that zinc deficiency increases the levels of inflammatory cytokines and oxidative stress, and induces apoptosis and endothelial cell dysfunction. Zinc supplementation reverses these adverse effects. Our recent *cell culture* studies have also demonstrated that zinc decreases the productions of oxidative stress and inflammatory cytokines, vascular cell adhesion molecule-1 (VCAM-1), as well as oxLDL-, and/or LPS-induced NF-κB activation in human monocyte-like and vascular endothelial cells. Our normal human subject study shows that zinc supplementation decreases the productions of oxidative stress biomarkers, TNF-α and IL-1β mRNAs, as well as TNF-α-induced NF-κB activation in isolated peripheral blood mononuclear cells (PMNC), compared to the placebo group. Normal elderly subjects show decreased levels of plasma and lymphocyte zinc, and increased generation of inflammatory cytokines and adhesion molecules in isolated PMNC cells, compared to normal young adult. Six months of zinc supplementation in normal elderly subjects increased plasma zinc and anti-oxidant capacity, and decreased plasma levels of C-reactive protein (CRP), IL-6, VCAM-1, MCP-1, and oxidative stress biomarkers, compared to the placebo-supplemented elderly subjects; this is consistent with the decreased incidence of infection. The data confirmed that zinc decreases the productions of inflammatory cytokines, oxidative stress, and atherosclerosis-associated biomarkers, suggesting that zinc may play a protective role in atherosclerosis through inhibition of inflammation and oxidative stress. In this chapter, we will discuss, in-depth, the molecular mechanisms of zinc action on the development and progression of atherosclerosis.

Chapter 5 - Zinc has recently emerged as an important trace element, capable of conferring cardioprotection and a variety of other benefits in the cardiovascular system and elsewhere. The events that occur subsequent to coronary artery occlusion are termed myocardial ischemia/reperfusion denoting that these are distinct phases of cellular injury with ATP depletion, lactate accumulation and acidosis observed during ischemia and production of reactive oxygen species during reperfusion. Cell death during myocardial ischemia/reperfusion occurs by necrosis and apoptosis, both regulated by stress signals and results in "cardiac wasting." Although the mechanisms by which zinc accomplishes cardiac benefits remain largely unclear, proposals include both direct and indirect effects on an array of intracellular targets such as adenosine receptor, ErbB2, NADPH oxidase, PKC, Akt, GSK-3β and mPTP culminating in anti-apoptotic effects and improved functional recovery of contractile function. Interestingly, cardioprotection by zinc is manifested only in the presence of a lipophilic ionophore, such as pyrithione, in isolated neonatal and adult rat cardiac myocytes subjected to simulated ischemia /reperfusion, Langendorff isolated rat heart perfusion and ischemia/reperfusion injury in myocardial infarction. Thus, although great potential exists for zinc-based therapeutics, a number of questions remain. This review will discuss the available literature pertaining to zinc homeostasis in the myocardium and provide novel insights into the underlying molecular mechanisms towards better understanding of the therapeutic effect of zinc during ischemia /reperfusion. Attempts are made to identify the gaps in basic and translational science in the development of zinc ionophores as potential therapeutic compounds in the setting of myocardial ischemia reperfusion injury.

Chapter 6 - The trace element zinc is ubiquitous in the human body. It has catalytic, structural or regulatory functions in more than 300 enzymes. The importance of an adequate zinc intake is marked by its impact on biological functions such as DNA replication, RNA transcription, signal transduction and redox regulation. In particular, cells of the innate and adaptive immune system depend on zinc ions, e.g., for proliferation, differentiation, and the production of cytokines. Hence, a deficiency affects the susceptibility to infections and the development of inflammation and autoimmunity. Zinc deficiency is mainly caused by malnutrition, but can also occur as a result of genetic defects of zinc transport proteins (e.g., in *Acrodermatitis enteropathica*) or secondary to diseases such as type 2 diabetes or diarrhea. Although a variety of studies have been performed regarding the therapeutic usefulness of zinc supplementation, the effects on several diseases remain unclear. Contradictory outcomes between different studies are caused by administration of diverse zinc salts and their varying bioavailability, the problematic determination of the patients' zinc status by plasma or serum zinc level, and differences in dosage and duration of the supplementation. For example, zinc is only effective in the treatment of the common cold if given within 24h of the onset of symptoms. Furthermore, zinc supplementation was shown to be beneficial for decreasing the frequency of diarrhea and pneumonia, diseases which contribute significantly to the high infant mortality in developing countries. On the other hand, zinc excess may result in increased susceptibility to infections by inhibiting leukocyte function. This review gives an overview on populations at risk of zinc deficiency, available supplements, and the impact of zinc administration on immune cells and several diseases.

Chapter 7 - Zinc is known as an essential nutritional factor in the growth of the human and animals. Bone growth retardation is a common finding in various conditions associated with dietary zinc deficiency. Bone zinc content has been shown to decrease in aging, skeletal unloading and postmenopausal conditions, suggesting its role in bone disorder. Zinc has been demonstrated to have a stimulatory effect on osteoblastic bone formation and mineralization; the metal directly activates aminoacyl-tRNA synthetase, a rate-limiting enzyme at translational process of protein synthesis, in the cells, and it stimulates cellular protein synthesis. Zinc has been shown to stimulate gene expression of the transcription factors runt-related transcription factor 2 (Runx 2) that is related to differentiation into osteoblastic cells. Moreover, zinc has been shown to inhibit osteoclastic bone resorption due to inhibiting osteoclast-like cell formation from bone marrow cells and stimulating apoptotic cell death of mature osteoclasts. Zinc has a suppressive effect on the receptor activator of nuclear factor (NF)-κB ligand (RANKL)-induced osteoclastogenesis. Zinc transporter has been shown to express in osteoblastic cells and osteoclastic cells. Zinc protein is invloved in transcription. The intake of dietary zinc causes an increase in bone mass. β-Alanyl-L-histidinato zinc (AHZ) is a zinc compound, in which zinc is chelated to β-alanyl-L-histidine. The stimulatory effect of AHZ on bone formation is more intensive than that of zinc sulfate. Zinc acexamate has also been shown to have a potent-anabolic effect on bone. The oral administration of AHZ or zinc acexamate has the restorative effect on bone loss under various pathophysiological conditions including aging, skeletal unloading, aluminium bone toxicity, calcium- and vitamin D-deficiency, adjuvant arthritis, estrogen deficiency, diabetes, and fracture healing. Zinc compounds may be designed as new supplymentation factor in the prevention and therapy of osteoporosis.

In: Zinc: Characteristics, Uses and Benefits
Editor: Masayoshi Yamaguchi
ISBN: 978-1-61942-523-1

Chapter 1

Inhibition Characteristic of Vegetable Extracts on the Corrosion of Al-Zn Alloys in Simulated Acidic Environment

***C. E. Ekuma,*[*1] *N. E. Idenyi,*[2] *I. O. Owate,*[1] *E. O. Chukwuocha,*[1] *E. Osarulobe*[1] *and M. C. Onyeaju*[1]**

[1] Department of Physics, University of Port Harcourt, Rivers, Nigeria.
[2] Department of Industrial Physics, Ebonyi State University, Abakaliki, Nigeria.

Abstract

A study of the effects of Inhibition Characteristic of Vegetable Extracts on the corrosion susceptibility of locally processed Al-base alloy in some media concentrations with varying volume of vegetable extracts (*Telferia occidentalis* and *Piper guinenses*) have been carried out. Virgin aluminum alloys containing 1.0% and 2.0% zinc were cast and machined to cylindrical coupons of initial surface area averaging $11.64cm^2$. The pre-weighed samples were carefully immersed into 0.5 M and 1.0 M of the acid containing 50 cm^3 and 100 cm^3 of the extracts, respectively. The set-ups were allowed to stand for 28 days with a set withdrawn weekly for corrosion rate characterization. The trend of the results obtained were

[*] E-mail address: panaceamee@yahoo.com

in tandem with that of passivating metals (or alloys) in simulated environments of initial steep rise in corrosion rate accompanied by a gradual decline as the exposure time increased. The vegetable extracts showed inhibition potentials with pronounced effects at higher extract concentrations. The best inhibiting property is found in Telferia occidentalis then, Piper guinenses.

INTRODUCTION

There has been recent increase in the study of the properties of aluminum based alloys and other low-density metals. This is noticeably due to their specific strength as engineering materials for transportation with relatively reduced fuel consumption and high strength – to – weight ratio [1]. For instance, a generation of aluminum-lithium alloys has been developed recently for use by aircraft and aerospace industries [2].

Aluminum owes its excellent corrosion resistance and its usage as one of the primary metals of commerce to the barrier oxide film, alumina (Al_2O_3) that shields it from corrosion attack. Aluminum has a strong affinity for oxygen, and in environments harboring aluminum (and some of its alloys), assists in the formation of this passive surface film bonded strongly to its surface, which is stable in aqueous media when the pH is between about 4.0 and 8.5, and if damaged, re-forms immediately in most environments [3]. The service life of most engineering materials depends on their ability to resist degradation. Comparatively, some metals are more intrinsically resistant to corrosion than others, either due to the fundamental nature of the electrochemical processes involved or due to the details of how reaction products form [4]. The complexity of the corrosion process may be better appreciated by recognizing that many variables are involved, including environmental, electrochemical and metallurgical aspects. For example, anodic reactions and rate of oxidation, cathodic reaction and rate of reduction, corrosion inhibition, polarization or retardation; passivity, effect of oxidizers, effect of velocity, temperature, corrosive concentration, galvanic coupling and metallurgical structure all influence the type and rate of the corrosion process [5].

Amongst the various corrosion control and prevention techniques, certain types of chemicals known as inhibitors may be added to the environment to reduce the aggressiveness of the environment or to reduce the corrosion process. Inhibitors function by ion- or molecule- adsorption onto a metal surface. The corrosion rate is reduced by increasing or decreasing the anodic and/or cathodic reaction, decreasing the diffusion rate for reactants to the surface of the metal, or increasing the electrical resistance of the metal surface [6]. However, there is a limitation to the type of environment that can be modified. The type of environment that can be modified include aqueous, partly aqueous (often found in at near neutral pH range such as natural waters, cooling water systems and in acidic range such as acid pickling to remove rolling scale as well as in the production and refinery of oil and gas) and gaseous [7].

The durability of metals or alloys in an aggressive environment depends on the protective properties of organic or inorganic films as well as nature of corrosion products. Some film characteristics, such as chemical composition, conductivity, adhesion, solubility, hygroscopy, and morphology, determine its ability as controlling barrier to different kinds of attack and corrosion rates [8]. On the other hand, these characteristics depend, in turn, on chemical composition and metallurgical history of the metal, on physicochemical properties of coating, and on environmental variables such as atmospheric conditions, type and amount of pollutants, wet-dry cycles, etc. [8,9,10].

The use of natural herbs: thyme, coriander, hibiscus, anis, black cumin and garden cress as a new type of green inhibitors acting on corrosion of steel has been investigated [11]. Yin has investigated the inhibition properties of natural honey in diverse environments [7].

Present chapter aims at investigating the service performance of Al-Zn alloys in various concentrations of hydrochloric acid containing varying concentration by volume of vegetable extracts: 50 cm^3 and 100 cm^3 extracts of *Telferia occidentalis* and *Piper guinenses*, respectively. The outcome would ensure proper understanding of the corrosion behavior of Al-Zn alloys when subjected to environments that are positively acidic, and hence in material selection during the design of engineering systems. Also, this is necessitated by the increasing concern about the toxic nature of some corrosion inhibitors presently in use in industry, which may not only affect living organisms but also bomee poison to the earth. It must be asserted that corrosion resistant alloys or sophisticated surface treatment is of little importance if parts are designed without the utilization of some basic corrosion knowledge. It will also be the cheapest and the best if potential corrosion cases can be eliminated at the design stage instead of dealing with the symptoms of bad design afterwards. Hence, the adoption of good professional ethics in material selection must be adhered to right at the design stage, construction stage and the service stage [12], i.e. to design-out corrosion out of the equipment [13].

The need for the present chapter is obvious. Al – Zn alloys have diverse uses in industries and in different environments. Material selection for engineering applications has been hindered by the deterioration of their physio-mechanical properties by large number of diverse environments. Even with the broad range of engineering materials modern technology has at its disposal, material selection for a given application has always been an arduous task. However, no general rule governs this choice, but a commonsensical decision normally involves a consideration of those properties that are to be conserved in any environment. Despite the cost of high quality, a balance must be struck between cost of frequent replacement/repairs and the ab-initio use of high quality materials [14]. This brings to fore, the dominant role played by corrosion in material selection involving metals which are naturally the most susceptible of all engineering materials to the deleterious effects of corrosion [15].

METHODOLOGY

Weight-Loss Technique

The weight loss technique is a simple, yet a robust method of evaluating the corrosion susceptibility in materials. This involves the total immersion of the pre-weighed sample tests in simulated environment for a given time and then, re-weighed to get the weight difference. The weight difference is then used to calculate the corrosion penetration rate of each of the samples. The corrosion penetration rate (CPR) or the rate of material removal as the consequence of the physiochemical interactions mostly expressed in mils/year or mm/yr is an important corrosion-monitoring index that gives reliable information to corrosion experts on the degree of corrosion progress in a given material in service. The rate of corrosion due to direct attack can be estimated from relatively simple laboratory test in which sample specimens of the related material are exposed to well-simulated actual environment with frequent weight change and dimensional measurements carefully taken. The mathematical computation of CPR, normally computed in mm / yr is based on the formula

$$CPR = \frac{87.6\,\Delta w}{\rho A t} \tag{1}$$

where Δw is weight loss in milligrams, t is the exposure time in second, ρ is the density of the specimen in grams per cubic centimeter, and A is the exposed specimen area of the specimen in square centimeter [2,16]. For most applications, a corrosion penetration rate of less than about 0.50mm/yr is acceptable [17]. However, this estimation is suitable only for uniform corrosion and becomes somewhat erratic with local instability [2].

For a passivated film to be effective, it must provide a protective barrier that keeps the corrosion current on the metal-environment interface at relatively low enough rate so that the extent of corrosion damage is minimized. An effective film is one that resists the breakdown of the passive film and if abraded, repassivates immediately [4,18]; hence, the conditions that promote corrosion of aluminum and its alloys, therefore, must be those that continuously abrade the film mechanically or promote reactions that locally degrade this protective oxide film by minimizing the availability of oxygen to rebuild it [5]. It should be noted that all breakdown mechanisms involve damaging species. Even at that, aluminum has excellent corrosion resistance, a consequence of barrier oxide film-the alumina (Al_2O_3) and this is the reason for its usage as one of the primary metals of commerce. Aluminum has a strong affinity for oxygen, and in environments harboring aluminum (and some of its alloys), assists in the formation of this passive surface film bonded strongly to its surface, which is stable in aqueous media when the pH is between about 4.0 and 8.5, and if damaged, re-forms immediately in most environments [12].

Materials / Equipment

The materials used for this work were scraps of virgin aluminum (99% pure Al) procured from aluminum stockist and pure granulated zinc metal procured also from metal stockist in Delta state, Nigeria. The other materials used were: *Piper guinenses* (Climbing-black pepper) and *Telferia occidentalis* (Fluted-pumpkin) which were sourced locally, acetone, hydrochloric acid (36.5% pure assay), distilled water, laboratory beakers, measuring cylinders, etc. The equipment used were: lathe machine, drilling machine, a surface crucible furnace and analytic digital weighing machine which measures to an accuracy of 0.000001 mg. This was used to weigh the test samples before and after immersion to know the exact weight difference.

Preparation of Test Samples

The aluminum scraps and zinc ingot were thoroughly cleaned to avoid contamination of its purity and possible degradation of its casting properties. The composition of each of the Al – Zn alloys were carefully worked out and charged into the surface crucible furnace. After melt down (□ $700^{o}C$), they were cast into cylindrical rods, machined to sizeable dimensions and consequently, cut into test coupons of initial specific surface area of about 11.64cm^2. Each sample coupon was drilled with a 5mm drill bit to provide hole for the suspension of the strings and the surface of each thoroughly polished according to ASTM standard. The samples for 1% Zn-99% Al alloy were coded "A" and 2% Zn-98% Al alloy were coded "B".

Simulation of Environment

The environments for this work were acidic environments of two concentrations (0.5M and 1.0M) made from 36.5% purity assay (the stock solution of hydrochloric acid) using standard procedure.

Preparation of the Extracts

Each of the vegetables: Fluted-pumpkin (*Telferia occidentalis*) and Climbing-black pepper (*Piper guinenses*) leaves were sourced locally. They are prepared by weighing 100g of the leaf using weighing balance. The weighed samples were minced in distilled water for reasonable period of time (about 20 minutes) to enable the green part of the leaf to be extracted and then filtered through glass wool. The filtrates were subsequently made up to 1.0 dm^3 of the solution using distilled water.

Design Setup and Experimentation

To 1.0 dm^3 of concentrations of the hydrochloric acid (0.5 M and 1.0 M) were added 50 cm^3 and 100 cm^3 of the *Piper guinenses* and *Telferia occidentalis* extracts, respectively. The extracts were found to be totally soluble in the acid. The media were carefully measured into eighteen (18) beakers, each containing four (4) test coupons of the alloys. A set of coupon was withdrawn 168 hourly (every 7 days), washed with distilled water, cleaned with acetone and dried in open air. The final weight of each of the test coupon was determined using the digital analytic weighing balance with an accuracy of about 0.000001 mg. In all, the samples were allowed to stand for 672 hours (28-days) with a set of test coupons withdrawn 168 hourly for corrosion rate characterization.

RESULTS

Tables 1 – 4 represents the corrosion rate data for Al – 1.0% Zn alloy and Al – 2.0% Zn alloy, respectively, in $50cm^3$ and $100cm^3$ volume concentrations of the vegetable extracts in 0.5 M and 1.0 M HCl concentrations of the simulated acidic environment.

A cursory look at the corrosion rate parameters (tables 1 – 3), show that the corrosion trend for most passivating metals was prominent in all the media studied. There was an initial steep rise in corrosion rate which gradually declined with increasing exposure time due to immobility of current carrying ions. That being the case, however, the degree of passivity varied significantly as the concentrations of the vegetable extracts changed. It will be observed that the degree of passivity varied albeit significantly with increasing concentrations of the acid. The observable differences in the inhibiting characteristics of the various vegetable extracts under the same acidic molarity may be attributed to their chemical composition. Also, the corrosion profile trends depict that at constant extract volume, the corrosion rates increased as the molarity of the acid increased. The explanation for this is that at constant extract volumes, the molarity of acid turns out to be a determining factor in the corrosion trend; thus, the higher the acid molarities, the higher the corrosion penetration rates. The corrosion rate values were also found to increase with increase in the reinforcing phase (here zinc) which is consistent in all the media studied. The implication of this is that the inhibition potentials of the extracts diminished with increase in the reinforcing phase except for *Piper guinenses*. In effect, the quantitative and chemical characteristics of the vegetable extracts influenced their inhibition potentials such that *Telferia occidentalis* proved to have a better inhibition to corrosion of Al-Zn alloys in simulated acidic environment than *Piper guinenses*. Consequently, a conclusion is drawn to the effect that the varying degree of inhibition potentialities of the vegetable extracts is a function of their individual chemical constituents as opposed to their volumetric characteristics in the acid media.

Table 1. Corrosion Rate Parameters of Al – 1.0% Zn immersed in 50cm^3 of the vegetable extracts in molar concentrations of HCl

Code	Time (Hours)	Initial Weight (g)	Final Weight (g)	**Weight Loss (g)**	**Corrosion rate (mm/yr)**
0.5 M HCl + 50cm^3 *Telferia occidentalis*					
A_1	168	10.1250	9.8012	0.3238	5.3537
A_2	336	9.9876	9.5287	0.4589	3.7937
A_3	504	10.3326	10.1988	0.1338	0.7374
A_4	672	10.8790	10.8685	0.0105	0.0434
0.5M HCl + 50cm^3 *Piper guinenses*					
A_5	168	10.1187	9.0422	1.0765	17.7987
A_6	336	10.8764	9.4274	1.4490	11.9788
A_7	504	10.5420	9.7456	0.7964	4.3892
A_8	672	10.8713	10.3181	0.5532	2.2866
1.0M HCl + 50cm^3 *Telferia occidentalis*					
A_9	168	11.3217	10.6826	0.6391	10.5668
A_{10}	336	11.0875	10.3927	0.6948	5.7439
A_{11}	504	10.9856	10.4126	0.5730	3.1580
A_{12}	672	10.7632	10.3201	0.4431	1.8315
1.0M HCl + 50cm^3 *Piper guinenses*					
A_{13}	168	11.1087	9.9767	1.1320	18.7163
A_{14}	336	11.4198	9.8262	1.5936	13.1742
A_{15}	504	10.8753	9.6359	1.2394	6.8307
A_{16}	672	10.9632	10.2983	0.6649	2.7483

Table 2. Corrosion Rate Parameters of Al – 2.0% Zn immersed in 50cm^3 of the vegetable extracts in molar concentrations of HCl

Code	Time (Hours)	Initial Weight (g)	Final Weight (g)	**Weight Loss (g)**	**Corrosion rate (mm/yr)**
0.5M HCl + 50cm^3 *Telferia occidentalis*					
B_1	168	10.1763	8.8354	1.3409	22.1702
B_2	336	10.2353	8.5561	1.6792	13.8818
B_3	504	10.3679	8.8847	1.4832	8.1743
B_4	672	10.6532	9.3002	1.3530	5.5926
0.5M HCl + 50cm^3 *Piper guinenses*					
B_5	168	10.3298	9.5255	0.8043	13.2982
B_6	336	11.0863	9.9878	1.0985	9.0812
B_7	504	10.3781	9.5916	0.7865	4.3346
B_8	672	11.2463	10.7991	0.4472	1.8485
1.0M HCl + 50cm^3 *Telferia occidentalis*					

Code	Time (Hours)	Initial Weight (g)	Final Weight (g)	**Weight Loss (g)**	**Corrosion rate (mm/yr)**
B_9	168	10.4396	8.6648	1.7748	29.3442
B_{10}	336	10.7923	8.9764	1.8159	15.0119
B_{11}	504	10.0958	8.7711	1.3247	7.3008
B_{12}	672	11.0174	10.0409	0.9765	4.0363
1.0M HCl + 50cm^3 *Piper guinenses*					
B_{13}	168	10.4396	8.6742	1.7654	29.1888
B_{14}	336	10.3958	8.5415	1.8543	15.3293
B_{15}	504	10.8021	9.4223	1.3798	7.6045
B_{16}	672	10.4089	9.4224	0.9865	4.0777

Table 3. Corrosion characteristics of Al – 1.0% Zn immersed in 100cm^3 of the vegetable extracts in molar concentrations of HCl

Code	Time (Hours)	Initial Weight (g)	Final Weight (g)	**Weight Loss (g)**	**Corrosion rate (mm/yr)**
0.5 M HCl + 100cm^3 *Telferia occidentalis*					
A_{17}	168	10.1879	9.9010	0.2869	4.7436
A_{18}	336	10.6532	10.2700	0.3832	3.1679
A_{19}	504	10.7653	10.5222	0.2431	1.3398
A_{20}	672	10.7301	10.5394	0.1907	0.7883
0.5M HCl + 100cm^3 *Piper guinenses*					
A_{21}	168	10.7620	10.2961	0.4659	7.7031
A_{22}	336	11.0988	10.5597	0.5391	4.4567
A_{23}	504	11.4322	11.0402	0.3920	2.1604
A_{24}	672	10.9539	10.7100	0.2439	1.0081
1.0M HCl + 100cm^3 *Telferia occidentalis*					
A_{25}	168	10.1542	9.6103	0.5439	8.9927
A_{26}	336	10.9759	10.3639	0.6120	5.0594
A_{27}	504	10.6523	10.2213	0.4310	2.3754
A_{28}	672	11.0958	10.6660	0.4298	1.7766
1.0M HCl + 50cm^3 *Piper guinenses*					
A_{29}	168	10.2301	8.7547	1.4754	24.3940
A_{30}	336	10.2319	8.8515	1.3804	11.4116
A_{31}	504	10.4309	9.3241	1.1068	6.0999
A_{32}	672	10.4386	9.4312	1.0074	4.1640

Table 4. Corrosion characteristics of Al – 2.0% Zn immersed in 100cm^3 of the vegetable extracts in molar concentrations of HCl

Code	Time (Hours)	Initial Weight (g)	Final Weight (g)	**Weight Loss (g)**	**Corrosion rate (mm/yr)**
0.5 M HCl + 100cm^3 *Telferia occidentalis*					
B_{17}	168	11.0236	10.4646	0.5590	9.2424
B_{18}	336	11.2190	10.5959	0.6231	5.1511
B_{19}	504	10.8743	10.3842	0.4901	2.7011
B_{20}	672	10.6592	10.3009	0.3583	1.4810

Table 4. (Continued)

Code	Time **(Hours)**	Initial Weight **(g)**	Final Weight **(g)**	**Weight Loss (g)**	**Corrosion rate (mm/yr)**
0.5M HCl + 100cm^3 *Piper guinenses*					
B_{21}	168	10.9037	9.1953	1.7084	28.2464
B_{22}	336	10.7329	8.9897	1.7432	14.4109
B_{23}	504	10.3820	8.7519	1.6301	8.9839
B_{24}	672	10.4821	8.9037	1.5784	6.5242
1.0M HCl + 100cm^3 *Telferia occidentalis*					
B_{25}	168	10.3291	8.4185	1.9106	31.5895
B_{26}	336	10.5302	8.2985	2.2317	18.4493
B_{27}	504	11.0934	9.4169	1.6765	9.2397
B_{28}	672	10.6539	9.0524	1.6015	6.6197
1.0M HCl + 50cm^3 *Piper guinenses*					
B_{29}	168	11.0385	8.3432	2.6953	44.5636
B_{30}	336	10.3941	7.4205	2.9736	24.5825
B_{31}	504	10.3095	7.6552	2.6543	14.6286
B_{32}	672	10.2149	8.4091	1.8058	7.4642

ACKNOWLEDGMENTS

The authors are grateful to the Management of Standards Organization of Nigeria, Delta State Laboratory for permission to use their facilities and to Eneh Esther for her contributions in this work. This work was funded in part by Ebonyi State, Federal Republic of Nigeria.

REFERENCES

[1] E. C. Ekuma, N. E. Idenyi, O. C. Obini, F. K. Onwu, and S. I. Neife, *R. J. Environ. Sci.* 1, 294 (2007).

[2] W. D. Callister, *Materials Science & Engineering: An Introduction, 4th Edition* (John Wiley and Sons Inc., New York, 1997), p. 520.

[3] E. C. Ekuma, E. Osarolube, E. O. Chukwuocha, M. C. Onyeaju, *W. J. Cond. Mat.* 1, 33 (2011).

[4] E. C. Ekuma, N. E. Idenyi, O. C. Obini, F. K. Onwu, and A. E. Umahi, *J. Appl. Sci.* 7, 237 (2007).

[5] N. E. Idenyi, E. C. Ekuma and I. O. Owate, *The Influence of Alloy Compositions on the Passivation Layer Characteristics of Al – Zn Alloys*

Systems. (The Impact of Design on Corrosion Performance, Material Science and Technology (MS&T) 2006) p. 613.

[6] E. C. Ekuma, N. E. Idenyi, G. O. Avwiri, and I. O. Owate, *Proceedings of Minerals, Metals and Materials (TMS) Annual Conference and Exhibition,* San Francisco, California, USA (15th – 19th Feb. 2009). Supplementary Proceedings 3, 2010) p. 51.

[7] J. Y. Yin, A thesis submitted to the University of Manchester Institute of Science and Technology for the degree of Master of Science, (2004).

[8] M. Stratmann, K. Bohnenkamp, and W. Engell, *Corrosion Science,* 23, 969 (1983).

[9] E. C. Ekuma and N. E. Idenyi, *Res. J. Phy.*, 1, 27 (2007).

[10] S.G. Stratmann and H. Streckel, *Corrosion. Sci.* 30, 697 (1990).

[11] E. Khamis and N. Al-Andis Mat.-wiss.u. Werkstofftech. 33, 550 (2002).

[12] E. C. Ekuma, N. E. Idenyi, and A. E. Umahi, *J. Appl. Sci.* 7, 237 (2007).

[13] M.N.C. Ijomah, *Elements of Corrosion and Prevention Theory,* (Enugu, Nigeria, Auto-Century Publishing Co. Ltd., 1991) p. 50.

[14] E. C. Ekuma, N. E. Idenyi, and S. I. Neife, *R. J. Environ. Sci.* 4, 185 (2007).

[15] E. C. Ekuma and N. E. Idenyi, *J. Appl. Sci.* 6, 1751 (2006).

[16] R. D. Joseph, ed. *Concise Metals Engineering Data Book* (ASM International U.S.A., 2007) p. 156.

[17] S. L. Kakani and A. Kakani, *Material Science* (New Age Publishers, India, 2006) p. 382.

[18] R. P. Frankenthal and J. Kruger, *Passivity of Metals* (The Electrochemical Society, Pennington, New Jersey, U.S.A. 1978).

In: Zinc: Characteristics, Uses and Benefits ISBN: 978-1-61942-523-1
Editor: Masayoshi Yamaguchi

Chapter 2

ZNO-BASED THICK-FILM ELECTRODOES FOR AN ANODE OF LI-ION BATTERIES

Hiroyuki Usui, Takamasa Kono and Hiroki Sakaguchi*

Department of Chemistry and Biotechnology, Graduate School of Engineering, Tottori University, 4-101 Minami, Koyama-cho, Tottori 680-8552, Japan.

ABSTRACT

A feasibility of ZnO-based materials was evaluated for a next-generation anode of Li-ion batteries. Composite thick-film electrodes of ZnO and Si were prepared by a gas-deposition method using an Si powder and a radial-shaped ZnO powder synthesized by a precipitation method. We evaluated anode performance of the composite electrodes, comparing with electrochemical properties of a ZnO electrode without Si. After the 20th cycle, the ZnO electrode exhibited a stable cyclability and a moderate discharge capacity of 160–220 mA h g^{-1}. The ZnO/Si composite electrode exhibited a better cycle stability after the 20th cycle and relatively-large discharge capacities exceeding the theoretical capacity of graphite in practical use until the 370th cycle. In contrast, the composite electrode containing of crushed ZnO showed flatter surface morphology and lower electrode performance in comparison with that using the radial-shaped ZnO. These results revealed that microstructure of the radial-shaped ZnO is very favorable for improving efficiencies of

* phone/fax: e-mail: sakaguch@chem.tottori-u.ac.jp

electrode reactions due to its high surface area, and is effective in accommodating the strains induced by Li-insertion and Li-extraction in/from Si.

INTRODUCTION

Next-generation anode materials with a high-specific capacity for Li-ion batteries are required so that electric vehicles widely spread. Silicon-based materials have recently attracted much attention as a promising candidate of the anodes due to their high-specific capacity in comparison with the practically-used graphite anodes. The specific capacity of Si is as high as 4200 mA h g^{-1}, which is more than ten times larger than that of 372 mA h g^{-1} in graphite anode. The remarkably large capacity is originated from an alloying reaction of Li–Si in the composition range from Si to $Li_{4.4}Si$ ($Li_{22}Si_5$) [1–3]. As additional advantages, Si is an abundant, less expensive, nontoxic element. Thus, Si-based materials are very appropriate as next-generation anodes. A practical use of Si for the anodes has been, however, disturbed by some critical disadvantages of its low intrinsic electrical conductivity, a low diffusion coefficient of Li in Si (10^{-14}–10^{-12} cm^2 s^{-1} [4–6]), and a drastic change of specific volume during the alloying/dealloying reactions. The volumetric change ratio per Si atom corresponds to be 410% [7–9], which causes a severe pulverization, an electrical isolation in the Si electrode, and a consequent capacity fading. Considerable attempts have been performed to overcome the problems. The attempts can be mainly classified in three ways. One is an alloying of Si and inactive elements such as transition metals to form silicides. When a silicide is used instead of elemental Si as the active material of the anode, the discharge capacity is drastically degraded owing to the smaller storage amount of Li ions in the silicide. We believe that anodes for next-generation Li-ion batteries should mainly consist of elemental Si to take an advantage of its huge theoretical capacity. The second way is structural modifications by utilizing amorphous Si [10] and nanostructured Si [11]. The third way is to form composites of active material (Si) and less-active materials such as carbon [12,13].

Gas-deposition (GD) method is a suitable for forming thick-film anodes and for improving their electrode performance, which has been recently revealed by the authors [14–21]. In this method, an aerosol consisting of raw particles and a carrier gas is jetted near the speed of sound from a nozzle onto a substrate. For preparing electrodes, we have demonstrated that the GD method has various advantages including (i) the strong adhesion between the active material particles as well as between the particles and the substrate, (ii) the nearly unchanging composition in the thick-film formed without atomization (*e.g.*, vaporization) of the particles, and (iii) the formation of interstitial spaces between particles, which is a favorable structure to release the stress induced by the volumetric change of the active material particles [14]. Furthermore, we fabricated composite thick-film electrodes consisting of elemental Si particles and other materials, and have discovered that the electrode performance is remarkably improved by the synergetic effects of the properties of Si and the combined materials, metals [17–19] and alloys [15,16,21]. However, the effect of metal oxides on the electrochemical properties of composite GD-film electrodes has never been explored though the metal oxides are also attractive anode materials on account of their large theoretical capacities.

Zinc oxide (ZnO) is one of the most versatile materials which have achieved applications in chemical sensors, piezoelectric devices, photocatalysts, photovoltaic cells, transparent electrodes, and luminescent materials of ultraviolet-visible lights. In addition to these applications, ZnO has been investigated as an active material of anode for Li-ion batteries [22–27] because ZnO is abundant, less expensive, and nonhazardous. It exhibits electrode reactions with a total theoretical capacity of 988 mA h g^{-1} as following equations:

$$ZnO + 2Li^+ + 2e^- \rightarrow Zn + Li_2O \quad (659\ mA\ h\ g^{-1},\ irreversible) \quad (1)$$

$$Zn + Li^+ + e^- \rightleftarrows LiZn \quad (329\ mA\ h\ g^{-1},\ reversible) \quad (2)$$

Volumes per mole of bulk ZnO, Zn, and Li_2O are 14.5, 18.2, and 14.8 cm^3, respectively. Consequently, a volumetric change from ZnO to Zn and Li_2O during Li insertion corresponds to 228%. in the equation (1). The significant volume expansion brings on pulverization and an electrical isolation in the active material, resulting in a serious capacity fade and a poor cycling performance. In addition, its low electronic conductivity and a slow kinetics of Li-ion diffusion in Zn [28–30] and ZnO [30] are also responsible for the poor performance. Some researchers have previously tried to improve the performance by reducing the size of ZnO. As for nano-sized ZnO electrodes [25,26], however, the performance has not been remarkably improved despite considerable efforts. To overcome the problem, the authors are herewith proposing GD-film electrodes consisted of microstructured ZnO prepared by a precipitation method, a novel one-step approach for chemical synthesis of ZnO. This method has been recently developed as a large-scale synthesis technique by some research groups, Yamabi *et al.* [31], Vayssieres *et al.* [32], and Li *et al.* [33]. The approach yields crystalline ZnO with a one-dimensional morphology in a strongly-basic aqueous solution containing $Zn(OH)_4^{2-}$ at mild conditions, atmospheric pressure and moderate temperatures below 90°C [31–35]. Following these reports, some researchers investigated a morphological control of ZnO while using this method. Those studies revealed that the ZnO microstructure can be easily controlled by varying growth conditions and by adding surfactants into the solutions [36–43].

In this study, we prepared thick-film electrodes consisted of a radial-shaped ZnO by using the GD method, and investigated their electrochemical Li-insertion/extraction properties as anodes of Li-ion batteries. For the purpose of a further enhancement of their electrode performance, ZnO/Si composite thick-film electrodes were also prepared by using mixed source materials of ZnO and Si. We evaluated the performance of the composite electrodes and investigated the influence of the ZnO microstructure on their cycling performance.

EXPERIMENTAL

Zinc oxide powders were synthesized by a precipitation method using strongly-basic aqueous solutions. The synthetic conditions in this study were determined by reference to other reports [31–33], but some procedures have been developed by the authors [41–43]: an aqueous solution of 0.31 mol/L zinc sulfate heptahydrate ($ZnSO_4{\bullet}7H_2O$) with the volume of 32.5 mL was firstly prepared for Zn^{2+} ion source. We made sodium hydroxide (NaOH) solution of 4.0 mol/L with 15 mL for pH adjustment, and sodium dodecyl sulfate (SDS, $C_{12}H_{25}SO_4Na$) solution of 0.2 mol/L with 2.5 mL. The solutions of $ZnSO_4{\bullet}7H_2O$ and SDS were poured into a flask, and agitated by a magnetic stirrer with a speed of 1000 rpm. The NaOH solution was added to the mixed solution. The pH value of the mixed solution was around 13. The mixed solution was stirred for 60 minutes at 5°C to obtain transparent solutions containing $Zn(OH)_4^{2-}$ ions [31–33], and that was subsequently stirred for 90 minutes at room temperature for nucleation of crystalline ZnO [31–33,43]. Then, hot stirring of the mixed solution was performed for 5 hours at 85°C to produce a white-colored precipitation. During the hot stirring, ZnO crystal growth gradually promoted by dehydration between OH^- on the surface of the growing crystals and the OH^- ligands of the hydroxyl complexes [31]. In the presence of SDS surfactant, the ZnO crystal surface was chemically activated by surfactant micelles, thereby forming a morphology of radial-shaped ZnO [41,43]. To remove surfactant molecules, the precipitation was centrifuged and washed with deionized water for three times. The washed precipitation was dried under a reduced pressure for 48 hours to obtain crystalline ZnO powder as a final product. All reagents were analytical grade and were used without further purification in the experiment. The crystal structure of ZnO was investigated by using an X-ray diffractometer (Ultima IV Rigaku Co., Ltd.) with CuKα radiation.

For gas-deposition, Cu foil substrates with 20 mm in thickness were set up in a guide tube with a nozzle of 0.8 mm diameter in a vacuum chamber [14–21]. The distance from the nozzle to the substrate was set to be 10 mm. An aerosol consisting of Ar gas (differential pressure: 7×10^5 Pa) and active material powders was generated in the guide tube, and instantly gushed from the nozzle onto the Cu substrate in the chamber with a base pressure of 10 Pa. As active material powders, we set the precipitated ZnO powder to prepare ZnO thick-films, or a mixture of the ZnO and a commercial Si powder (Wako Pure Chemical Industries, Ltd., 99.9%) to obtain ZnO/Si composite thick-films. To change the ratio of ZnO to Si in the obtained films, A blend ratio of ZnO:Si in the mixtures was varied. The weight ratios of ZnO/Si were 70/30, 75/25, and 89/11 wt.% in the thick-film electrodes, which was analyzed by means of an energy dispersive X-ray fluorescence (XRF) spectrometer (EDX-720 Shimadzu Co. Ltd.). The amount of the deposited active materials on the substrates was about 0.1 mg. The surface morphologies of the thick-films were observed using a field-emission scanning electron microscope (JSM-6701F JEOL Co., Ltd.).

Electrochemical measurements were carried out with a beaker-type three-electrode cell. The working electrodes were the obtained thick-film electrodes. Both counter and reference electrodes were 1-mm-thick Li metal sheets (Rare Metallic Co., Ltd., 99.90%). We used $LiClO_4$-dissolved in propylene carbonate (PC; $C_4H_6O_3$, Kishida Chemical Co., Ltd.) at a concentration of 1.0 M as the electrolyte. Constant current charge–discharge tests were performed using an electrochemical measurement system (HZ-3000 Hokuto Denko Co., Ltd.) under constant currents at 303 K with the cutoff potentials set as 0.005 V vs. Li/Li^+ for charge and 3.000 V vs. Li/Li^+ for discharge. The current densities were set to be 1.0 A g^{-1} (1.0*C*, 1*C* was defined as a current rate of three Li^+ reaction per hour; 988 mA g^{-1}) for ZnO electrodes, and 2.7–3.2 A g^{-1} (approximately 1.0*C*) for ZnO/Si composite electrodes. Cyclic voltammetry measurements were carried out using an electrochemical measurement system (Compactstat Ivium Technologies Co., Ltd.) at a scan rate of 0.1 mV s^{-1}.

Results and Discussion

Characterization of ZnO Electrode

Figure 1(a) displays a FE-SEM image of the obtained powder by the precipitation method in this study. Hexagonal columns with a diameter of 50–500 nm and a length of 1–5 mm were densely aggregated, forming a radial-shaped structure. The well-faceted surface of prism planes like a pencil was recognizable for the hexagonal columns. The inset in the figure is an enlarged view for a basal plane of one column. A regular hexagon can be confirmed, indicating the crystallization in the hexagonal system. Figure 1(b) shows a XRD pattern of the precipitated powder.

All diffraction peaks are very sharp and can be indexed as wurtzite ZnO (ICSD No. 00-36-1451). No diffraction peak of SDS surfactant and other materials was observed. These results revealed that the obtained powder consisted of well-crystallized ZnO with the radial-shaped morphology.

Figure 2 gives cyclic voltammograms of a ZnO thick-film electrode prepared by gas-deposition using a source material of the precipitated ZnO powder. Note that the sweep rate was set to be lower value of 0.1 mV s^{-1} than typical conditions to identify obtained peaks because the chemical diffusion coefficient of Li ion in ZnO is very low (10^{-14}–10^{-12} cm^2 s^{-1} [30]): Li diffusion in ZnO is very slow. At the first cycle, a large cathodic peak appeared at 0.47 V vs. Li/Li^+. This peak is attributed to a reduction reaction of ZnO in the equation (1) to form Li_2O. At the following cycles, the peaks continued to appear though the peak current density was significantly reduced. The reduction reaction of ZnO is suggested to take place at the following cycles. Below the potential of 0.20 V vs. Li/Li^+, broad cathodic peaks were observed at every cycle. These peaks are ascribed to four-stage alloying/dealloying reactions of Li–Zn system as follows:

$$Zn + 0.4Li^+ + 0.4e^- \rightleftarrows Li_{0.4}Zn \quad (3)$$

$$Li_{0.4}Zn + 0.1Li^+ + 0.1e^- \rightleftarrows Li_{0.5}Zn \quad (4)$$

$$Li_{0.5}Zn + 0.17Li^+ + 0.17e^- \rightleftarrows Li_{0.67}Zn \quad (5)$$

$$Li_{0.67}Zn + 0.33Li^+ + 0.33e^- \rightleftarrows LiZn \quad (6)$$

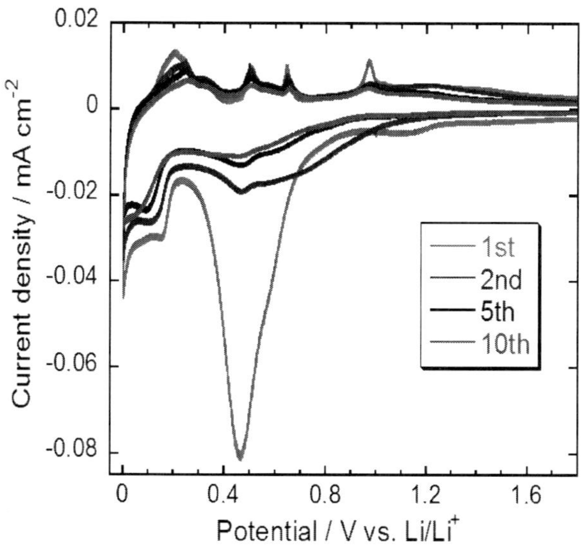

Figure 2. Cyclic voltammograms of ZnO thick-film electrode prepared by gas-deposition.

We consider that four kinds of original cathodic peaks were merged into one peak due to the very slow kinetics of the first lithiation stage from Zn to

$Li_{0.4}Zn$ [28–30], which has been confirmed by Fujieda *et al.* [29]. In the first cycle of the delithiation process, four anodic peaks were observed at 0.20, 0.31, 0.52, and 0.66 V vs. Li/Li^+, corresponding to the four-stage dealloying reactions of Li–Zn system described as the equations from (3) to (6).

The potentials of the four peaks were in good agreement with those in the previous reports [27,29,30]. Another peak located at 0.98 V vs. Li/Li^+ was presumably caused from a stripping of Li-underpotential deposition (UPD) on Cu substrate [44–46]. In addition, a very-broad anodic peak can be recognized in the potential range from 0.8 to 1.6 V vs. Li/Li^+, which is probably attributed to a reverse reaction of the equation (1), reduction of Li_2O and oxidation of Zn. This is one of the conversion reactions of metal oxides discovered by Tarascon *et al.* [47–50] and other groups [51–54]. In the subsequent cycles, the voltammogram profiles showed a good reproducibility though the peak intensities were reduced, indicating that the alloying/dealloying reactions of Li–Zn were reversible and that the conversion reaction of ZnO is partially reversible.

Figure 3(a) shows charge–discharge curves at the initial two cycles of the ZnO thick-film electrode. During the initial charge process, the electrode potential rapidly dropped to reach a plateau at 0.4–0.5 V vs. Li/Li^+, and then continuously decreased down to 0.2 V. The potential plateau corresponds to the formation of Zn and Li_2O. Another plateau in the potential range of 0–0.2 V vs. Li/Li^+ indicates the alloying reactions of Li–Zn as mentioned in the interpretation for the voltammograms. During the first discharge process, some gentle shoulders were recognized from 0 to 1.5 V vs. Li/Li^+.

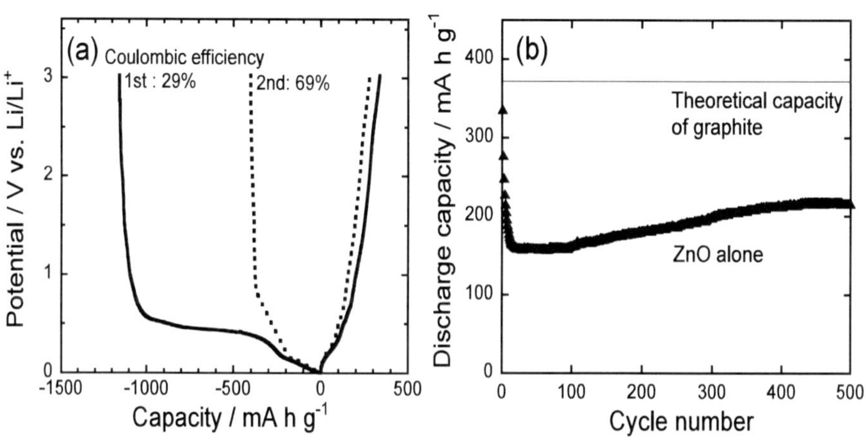

Figure 3. (a) Charge–discharge curves at the initial two cycles and (b) cycling performance of ZnO thick-film electrode.

According to the CV results, the dealloying of Li–Zn and the oxidation of Zn to ZnO are possibly responsible for the potential shoulders. At the first cycle, the charge (Li-insertion) and discharge (Li-extraction) capacities were 1140 and 340 mA h g^{-1}, resulting in a low coulombic efficiency of only 29%. The reason for the large irreversible capacity is that the formed Li_2O is not easily decomposed to the elemental Li in the discharge process owing to a high thermodynamic stability of Li_2O (standard enthalpy of formation: –599 kJ mol^{-1}) [55]. On the following cycles, similar charge–discharge curves were obtained though the plateau of the charge process at 0.4–0.5 V shortened and inclined. From these results, the ZnO electrode prepared in this study exhibited the characteristic electrochemical signature of the conversion reaction involved in transition-metal oxides [47].

Figure 3(b) depicts dependence of the discharge capacity on the charge–discharge cycle for the ZnO electrode. The capacity quickly diminished from 340 to 160 mA h g^{-1} in the initial 20 cycles, which is related to the formation of inactive Li_2O. In the subsequent cycles, the electrode exhibited very stable cycling performance and gradually increasing capacities though the capacities were lower than the theoretical capacity of graphite. In view of the stable cycling performance, ZnO obtained by the precipitation method would be a feasible material to be combined with Si as our composite GD-film electrode.

Electrochemical Properties of ZnO/Si Composite Electrodes

For ZnO/Si composite thick-film electrodes with different weight ratios of

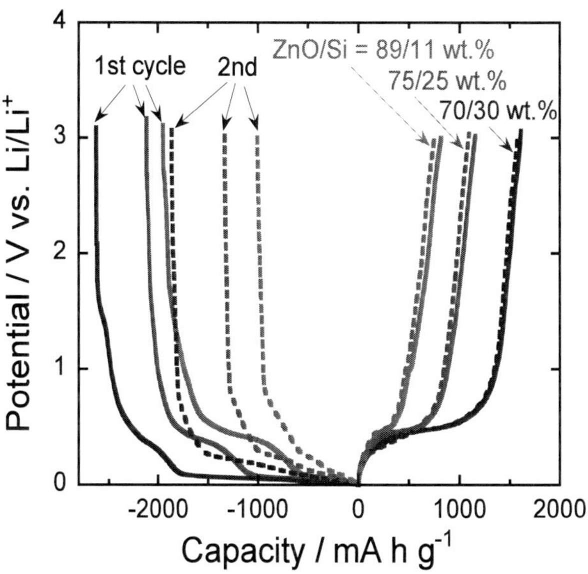

Figure 4. Charge–discharge curves at the initial two cycles of ZnO/Si composite thick-film electrodes with different weight ratios of ZnO and Si.

ZnO to Si, Fig.4 shows charge–discharge curves at the initial two cycles.

The three electrodes behaved very alike. At the first cycle, these composite electrodes also showed potential plateaus at 0.4–0.5 V originated from the formation of Zn and Li_2O as is the case with the ZnO electrode. In addition to these, we obviously confirmed plateaus in charge and discharge processes at 0.05 V and 0.46 V vs. Li/Li^+, respectively. These potential plateaus are attributed to the alloying/dealloying reactions of Si with Li. We have proposed that an active material, combined with Si in composite electrodes, should be prohibited from blocking out Li-ion storage in the active material to enable efficient Li-ion transportation into/from Si [15,16,21]. The ZnO has a moderate Li-storage ability; moderate discharge capacities of 160–220 mA h g^{-1} were achieved after the initial capacity decay as demonstrated in Fig. 3. Owing to the moderate Li-storage ability of ZnO, Li-insertion and Li-extraction into/from Si were performed in the ZnO/Si composite electrodes. The inclined plateaus in charge process at the second cycle are presumably attributed to amorphization of Si by the first Li-insertion/extraction. The charge and discharge capacities were reduced with decreasing the weight ratio of Si in the composite electrodes. We thus found that the Si amount involved in the electrodes basically contributes to their charge and discharge capacities because of much larger theoretical capacity of Si than that of ZnO.

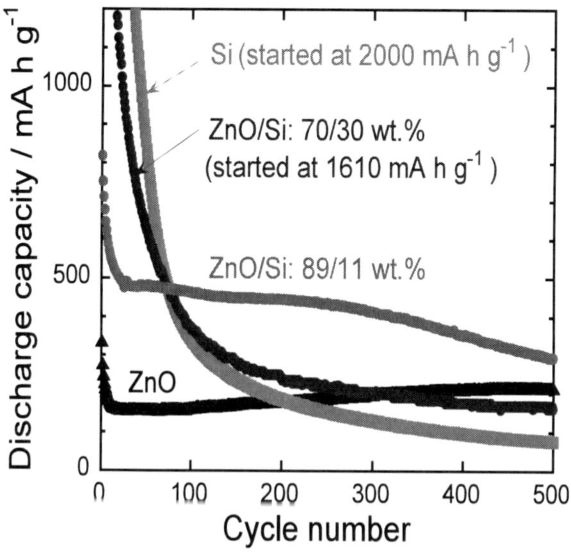

Figure 5. Dependence of discharge capacity on charge–discharge cycle for ZnO/Si composite thick-film electrodes.

Figure 5 summarizes cycling performance of these composite electrodes. For comparison, the data were also plotted in the figure for two electrodes consisted of Si alone and ZnO alone. A large discharge capacity of approximately 2000 mA h g^{-1} was obtained for the Si electrode. A steep decay of the discharge capacity was, however, observed until the 100th cycle, which is attributed to pulverization and an electrical isolation of Si induced by its drastic volumetric change during Li-insertion/extraction. A similar capacity decay was observed in case of the ZnO/Si composite electrode including Si of 30 wt.%. It is suggested that the relatively-large amount of Si brought on the pulverization and the electrical isolation of the composite electrode. On the other hand, the composite electrode containing Si of 11 wt.% showed a rapid capacity drop from 820 to 500 mA h g^{-1} in the initial 20 cycles. After the 20th cycle, a good cycle stability was achieved. This behavior closely resembles that of the ZnO electrode. It is a remarkable result that the capacity of the composite electrode exceeded the theoretical capacity of graphite until 370th cycle.

Influence of ZnO Microstructure on Composite Electrode Performance

To study the origin the better cycling performance of the composite electrode, we investigated the influence of ZnO microstructure on the electrode performance. Figure 6 compares cycling performance of two ZnO/Si composite electrodes. One is the above-mentioned ZnO/Si electrode with 89/11 in wt.%. The other is a composite electrode prepared by GD using a source material consisted of Si and a crushed ZnO powder by a milling process in an agate mortar to break up the unique microstructure of hexagonal ZnO columns. Its weight ratio was 90/10 wt.%, which is nearly the same as that of the electrode using pristine ZnO. The composite electrode using the crushed ZnO exhibited a low performance; a small initial capacity of 300 mA h g^{-1} was obtained and the capacity was continuously decreased. Obviously, on the other hand, the composite electrode containing pristine ZnO delivered much better electrode performance. Insets in Fig. 6 display FE-SEM images of these electrodes before the charge–discharge. In case of the electrode containing pristine ZnO, hexagonal ZnO columns were densely aggregated to form numerous interspaces on the surface. Such microstructure would be very favorable for improving efficiencies of electrode reactions due to its high surface area, and for enhancing charge/discharge capacities. In addition, the microstructure with the many interspaces can effectively accommodate the strains induced by Li-insertion/extraction. The accommodation suppresses the pulverization of the active materials, and thus improves the cycle stability of electrode. By contrast, we observed a flatter surface morphology for the electrode consisted of crushed ZnO. No interspace was found on the surface unlike the other electrode because the radial shape and hexagonal columns of ZnO were broken up by the milling process. Therefore, a lower surface area of the electrode limits the efficiency of electrode reactions to considerably reduce its discharge capacity. These results clearly revealed that the radial-shaped ZnO prepared by the precipitation method has the unique microstructure effective for improving the electrode performance of ZnO/Si composite electrodes because of the moderate Li-storage ability of ZnO and the numerous interspaces between the hexagonal ZnO columns.

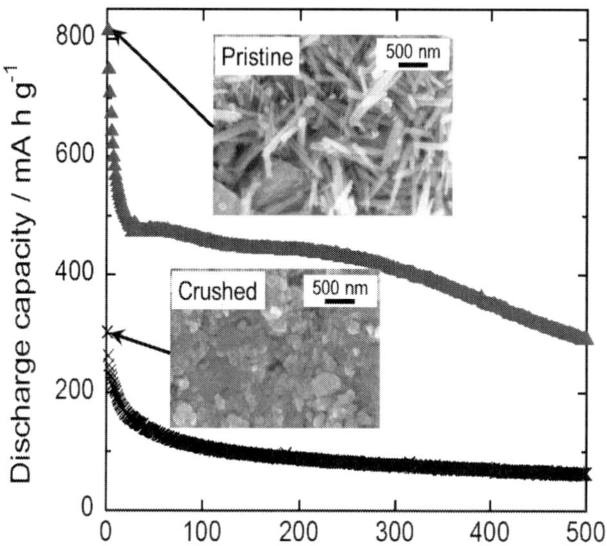

Figure 6. Influence of crushed ZnO on cycling performance of ZnO/Si composite thick-film electrodes.Figure 1. (a) FE-SEM image and (b) XRD pattern of Figure 1. precipitated powder for source material of gas-deposition.

CONCLUSION

The radial-shaped ZnO powder was synthesized by the precipitation method as the active material to be combined with Si in composite thick-film anode for next generation Li-ion batteries. The ZnO/Si composite thick-film electrodes were prepared by the gas-deposition method using the obtained ZnO and Si. We evaluated the anode performance for the ZnO/Si composite electrodes, comparing with the electrochemical properties of the ZnO electrode without Si. Two kinds of electrode reactions, the reversible alloying/dealloying reactions of Li–Zn and the partially-reversible conversion reaction of ZnO with Li, were confirmed by the CV measurements of the ZnO electrode. We observed that the ZnO electrode has the stable cycle stability and the moderate Li-storage ability. The composite electrode of ZnO/Si with 89/11 in wt.% exhibited the better cycle stability after the 20th cycle and the relatively-large discharge capacities exceeding the theoretical capacity of graphite until 370th cycle. In contrast, the composite electrode consisted of crushed ZnO showed the flatter surface morphology and the lower electrode performance in comparison with that using radial-shaped ZnO. These results revealed that the radial-shaped ZnO has a very favorable microstructure for improving efficiencies of electrode reactions due to its high surface area, and that the microstructure is effective in accommodating the strains induced by Li-insertion and Li-extraction in/from Si.

ACKNOWLEDGMENTS

This work was partially supported by a grant from the project, Development of High-performance Battery System for Next generation Vehicles (Li-EAD Project), from the New Energy and Industrial Technology Development Organization (NEDO) of Japan. This work was partly supported from the Grant-in-Aid for Scientific Research of Ministry of Education, Culture, Sports, Science and Technology (MEXT), Japan.

REFERENCES

[1] J. O. Besenhard. J. Yang, M. Winter, *J. Power Sources*, 68 (1997) 87.
[2] U. Kasavajjula, C. S. Wang, A. J. Appleby, *J. Power Sources*, 163 (2007) 1003.
[3] N. Ding, J. Xu, Y. Yao, G. Wegner, I. Lieberwirth, C. Chena, *J. Power Sources*, 192 (2009) 644.
[4] H. Xia, S. Tang, L. Lu, *Mater. Res. Bull.*, 42 (2007) 1301.

[5] N. Ding, J. Xu, Y. X. Yao, G. Wegner, X. Fang, C. H. Chen, I. Lieberwirth, *Solid State Ionics*, 180 (2009) 222.
[6] J. Xie, N. Imanishi, T. Zhang, A. Hirano, Y. Takeda, O. Yamamoto, *Mater. Chem. Phys.*, 120 (2010) 421.
[7] C. J. Wen, R. A. Huggins, *J. Solid State Chem.*, 37 (1981) 271.
[8] V. L. Chevrier, J. W. Zwanziger, J. R. Dahn, *J. Alloys Compd.*, 496 (2010) 25.
[9] H. Kim, C.-Y. Chou, J. G. Ekerdt, G. S. Hwang, *J. Phys. Chem. C*, 115 (2011) 2515.
[10] K.-L. Lee, J.-Y. Jung, S.-W. Lee, H.-S. Moon, J.-W. Park, *J. Power Sources*, 129 (2004) 270.
[11] M.-H. Park, M. G. Kim, J. Joo, K. Kim, J. Kim, S. Ahn, Y. Cui, J. Cho, *Nano Lett.*, 9 (2009) 3844.
[12] I.-S. Kim, P. N. Kumta, J. Power Sources, 136 (2004) 145.
[13] Y. Zhang, X. G. Zhang, H. L. Zhang, Z. G. Zhao, F. Li, C. Liu, H. M. Cheng, *Electrochim. Acta*, 51 (2006) 4994.
[14] H. Sakaguchi, T. Toda, Y. Nagao, T. Esaka, *Electrochem. Solid-State Lett.*, 10 (2007) J146.
[15] T. Iida, T. Hirono, N. Shibamura, H. Sakaguchi, *Electrochemistry*, 76 (2008) 644.
[16] H. Sakaguchi, T. Iida, M. Itoh, N. Shibamura, T. Hirono, *IOP Conf. Series: Mater. Sci. Eng.*, 1 (2009) 012030-1.
[17] H. Usui, Y. Kashiwa, T. Iida, H. Sakaguchi, *J. Power Sources*, 195 (2010) 3649.
[18] H. Usui, H. Nishinami, T. Iida, H. Sakaguchi, *Electrochemistry*, 78 (2010) 329.
[19] H. Usui, M. Shibata, K. Nakai, H. Sakaguchi, *J. Power Sources*, 196 (2011) 2143.
[20] H. Usui, Y. Yamamoto, K. Yoshiyama, T. Itoh, H. Sakaguchi, *J. Power Sources*, 196 (2011) 3911.
[21] H. Usui, K. Meabara, K. Nakai, H. Sakaguchi, *Int. J. Electrochem. Sci.*, 6 (2011) 2246.
[22] F. Belliard, P. A. Connor, J. T. S. Irvine, *Ionics*, 5 (1999) 450.
[23] F. Belliard, P. A. Connor, J. T. S. Irvine, *Solid State Ionics*, 135 (2000) 163.
[24] F. Belliard, J. T. S. Irvine, *J. Power Sources*, 97-98 (2001) 219.
[25] J. P. Liu, Y. Y. Li, X. T. Huang, G. Y. Li, Z. K. Li, *Adv. Funct. Mater.*, 18 (2008) 1448.

[26] J. Liu, Y. Li, R. Ding, J. Jiang, Y. Hu, X. Ji, Q. Chi, Z. Zhu, X. Huang, *J. Phys. Chem. C*, 113 (2009) 5336.
[27] H. Wang, Q. Pana, Y. Cheng, J. Zhao, G. Yin, *Electrochim. Acta*, 54 (2009) 2851.
[28] J. Wang, P. King, R. A. Huggins, *Solid State Ionics*, 20 (1986) 185.
[29] T. Fujieda, S. Takahashi, S. Higuchi, *J. Power Sources*, 40 (1992) 283.
[30] J. Xie, N. Imanishi, A. Hirano, Y. Takeda, O. Yamamoto, X. B. Zhao, G. S. Cao, *Thin Solid Films*, 519 (2011) 3373.
[31] S. Yamabi, H. Imai, *J. Mater. Chem.*, 12 (2002) 3773.
[32] L. Vayssieres, *Adv. Mater.*, 15 (2003) 464.
[33] P. Li, Y. Wei, H. Liu, X.K. Wang, *Chem. Commun.*, 24 (2004) 2856.
[34] W. W. He, Y. P. Li, Z. Q. Chen, Y. Q. Wang, R.Y . Zhang, *Mater. Lett.*, 60 (2006) 2299.
[35] J. Liu, X. Huang, Y. Li, J. Duan, H. Ai, *Mater. Chem. Phys.*, 98 (2006) 523.
[36] Y. Masuda, N. Kinoshita, K. Koumoto, *Electrochim. Acta*, 53 (2007) 171.
[37] P. Li, H. Liu, Y.F. Zhang, Y. Wei, X.K. Wang, *Mater. Chem. Phys.*, 106 (2007) 63.
[38] S. T. He, H. Maeda, M. Uehara, M. Miyazaki, *Mater. Lett.*, 61 (2007) 626.
[39] L. Tang, B. Zhou, Y. Tian, H. Bala, Y. Pan, S. Ren, Y. Wang, X. Lv, M. Li, Z. Wang, *Colloid Surf. A*, 296 (2007) 92.
[40] R. Wahab, S. G. Ansari, Y. S. Kim, H. K. Seo, G. S. Kim, G. Khang, H.-S. Shin, *Mater. Res. Bull.*, 42 (2007) 1640.
[41] H. Usui, *J. Phys. Chem. C*, 111 (2007) 9060.
[42] H. Usui, *Mater. Lett.*, 63 (2009) 1489.
[43] H. Usui, *J. Colloid Interface Sci.*, 336 (2009) 667.
[44] D. M. Kolb, M. Przasnyski, H Gerischer, *Electroanal. Chem. Interfacial Electrochem.*, 54 (1976) 25.
[45] L.-F. Li, D. Totir, Y. Gofer, G. S. Chottiner, D. A. Scherson, *Electrochim. Acta*, 44 (1998) 949.
[46] M. Moshkovich, Y. Gofer, D. Aurbach, *J. Electrochem. Soc.*, 148 (2001) E155.
[47] P. Poizot, S. Laruelle, S. Grugeon, L. Dupont, J.-M. Tarascon, *Nature*, 407 (2000) 496.
[48] S. Grugeon, S. Laruelle, R. H. Urbina, L. Dupont, P. Poizot, J.-M. Tarascon, *J. Electrochem. Soc.*, 149 (2001) A285.

[49] P. Poizot, S. Laruelle, S. Grugeon, J.-M. Tarascon, *J. Electrochem. Soc.*, 149 (2002) A1212.

[50] S. Mitra, P. Poizot, A. Finke, J.-M. Tarascon, *Adv. Funct. Mater.*, 16 (2006) 2281.

[51] J. Cabana , L. Monconduit , D. Larcher , M. R. Palacín, *Adv. Mater.*, 22 (2010) E170.

[52] P. Balaya, H. Li, L. Kienle, J. Maier, *Adv. Funct. Mater.*, 13 (2003) 621.

[53] O. Delmer, P. Balaya, L. Kienle, J. Maier, *Adv. Mater.*, 20 (2008) 501.

[54] E. Bekaert, P. Balaya, S. Murugavel, J. Maier, M. Ménétrier, *Chem. Mater.*, 21 (2009) 856.

[55] G. K. Johnson, R. T. Grow, W. N. Hubbard, *J. Chem. Thermodynamics*, 7 (1975) 781.

In: Zinc: Characteristics, Uses and Benefits ISBN: 978-1-61942-523-1
Editor: Masayoshi Yamaguchi

Chapter 3

ZINC IN PROSTATE CANCER: THE STUDY BASED ON SYNCHROTRON RADIATION TECHNIQUES

A. Banas[a,b*], K. Banas[a,b], W. M. Kwiatek[b], M. Gajda[c], B. Pawlicki[d] and T. Cichocki[c]

[a] Singapore Synchrotron Light Source (SSLS), National University of Singapore, 5 Research Link, Singapore 117603.
[b] Institute of Nuclear Physics, Polish Academy of Sciences, Radzikowskiego 152, 31-342 Kraków, Poland.
[c] Department of Histology, Collegium Medicum, Jagiellonian University, Kopernika 7, 31-034 Kraków, Poland [d]Gabriel Narutowicz Hospital, Prądnicka 37, 31-202 Kraków, Poland.

ABSTRACT

Zinc is a vital element in the nutrition of human beings, animals and plants. Although most experimental data have demonstrated the adverse effects of Zn deficiency on human and animal health, there is still evidence to support the hypothesis that Zn toxicity can also result in deleterious health effects. Conflicting views have been expressed regarding the role of Zn in carcinogenesis. Some studies emphasize the fact that Zn deficiency causes cancers, while others highlight that Zn is involved in tumour growth and development of neoplastic transformation.

* Corresponding author: Agnieszka Banas; e-mail address: slsba@nus.edu.sg

These conflicting observations suggest that the role of Zn in different organs may be different.

Based on the results of a micro-beam synchrotron X-ray fluorescence technique (micro-SRIXE) applied to the quantitative analysis of Zn, we correlated Zn concentration levels with the clinical stage of prostate cancer at the time of operation (described by the Gleason grade system). Serial sections of prostate tissues were collected from patients undergoing radical prostatectomy. Our results seem to be valuable in light of determination of the potential changes in Zn concentration as a diagnostic marker and its etiological involvement in the different stages of prostate diseases.

INTRODUCTION

Prostate Cancer

The prostate is a small, squishy gland the size of a walnut. It is situated under the bladder and in front of the rectum. Male hormones are responsible for the growth of the prostate gland in the foetus. The prostate gland grows until a man reaches adulthood and then stays about the same size as long as male hormones are produced [1].

Prostate cancer is the most frequently diagnosed cancer in men and is the second leading cause of cancer death [2]. Intensified screening efforts in recent years and the use of prostate-specific antigen (PSA) as a marker of disease have resulted in increased diagnoses in the early stages of the disease. While researchers still do not know exactly what causes prostate cancer, they have identified some factors that increase the risk of getting it including age, family history, ethnicity, diet and obesity [1-3]. Prostate cancer occurs when cells in the prostate gland grow and multiply uncontrollably. The nodular tumour mass invades the surrounding tissues and eventually spread to the lymph nodes, skeletal bones and other regions of the body. Although several cell types are found in the prostate, over 99% of prostate cancers develop from the glandular cells and is therefore described as *Adenocarcinoma* [3, 4].

Over the past fifteen years, more detailed descriptions of the cancers have been given by pathologists; prostate tissues were classified using a new tool - Gleason grade [5-7]. This system is based on the architectural patterns of cancer glands observed by the pathologist under the microscope (Fig. 1). Whether or not the cancer is likely to grow slowly or very quickly depends on, according to the Gleason grading, how closely the cells of the cancer resemble normal cells. These cells that look like normal glandular tissue with some cancer cells in the lining are termed as well-differentiated cancers. They have a low Gleason grade 1 or 2, grow slowly and take a long time to become clinically significant.

Cells that appear very little like normal glands and more like sheets of packed cancer cells are termed poorly-differentiated cancers (Gleason grade 4 or 5) and these tumours are very aggressive and have poor results with most treatments.

Finally, cells which form multiple small glandular spaces are in the middle of the Gleason grade scale and are called moderately-well- differentiated cancers (Gleason grade 3).

Undoubtedly Gleason grading provides valuable information in terms of prostate cancer classification, but limitations in this technique include disparity between tissue architecture and its biochemical status. Understanding what exactly happens when prostate cancer begins to grow is very important. It may also help in the selection of reliable treatment options and it could help in saving many men's lives.

There is still a need for additional markers which could increase sensitivity and specificity of prostate cancer detection and also the selection therapy for the treatment of disease. Some authorities in urology have suggested that all new prognostic markers should be evaluated in conjunction with the Gleason grade [8].

Zinc

For many years, studies of concentration of trace elements have played an important role in the field of better understanding the different processes occurring in cells. Instability (overabundance or deficiency) in the composition of trace metals that are recognized to be essential to normal human homeostasis and the accumulation of potentially toxic or nonessential trace metals may lead to different diseases. It was also proven that some trace elements play critical roles in cancer biology; however, there is still a serious lack of information concerning their function in initiation, promotion, and inhibition of carcinogenic processes in prostate glands [9-15]. Thus, there is a real need for analysis of trace elements in cancerous and noncancerous human tissues to examine the relationship between cancer and these elements. Detailed *in situ* studies are necessary if the role of metal accumulation in disease pathology is to be understood.

More than 100 years ago, zinc was recognized as a trace element with important roles in various metabolic processes in living organisms [16]. Being the second most abundant transitional metal ion in the human body, zinc is essential for the proper function of different enzymes and for a tight control of gene expression [17]. Zn is treated also as an essential element for the immune system [18, 19]. Its concentration level has to be balanced for the proper functioning of the immune system since both increased and decreased amounts of this element result in a disturbed immune function. Alterations in immune function increase the risk of several diseases including cancer.

Although most experimental data have demonstrated the adverse effects of Zn deficiency on human and animal health, there is still evidence to support the hypothesis that Zn toxicity can also result in deleterious health effects [20-22]. Conflicting views have been expressed regarding the role of Zn in carcinogenesis. Some studies [23, 24] emphasize the fact that Zn deficiency causes cancers while others [22, 25, 26] highlight that Zn is involved in tumour growth and development of neoplastic transformation. Disturbed zinc homeostasis has been reported in many types of cancer, including breast, prostate, lung, and brain [10, 27, 28].

These conflicting observations suggest that the role of Zn in different organs may be different. The zinc concentration level in normal prostate tissue is about 150 mg/g wet weight, which is 3–5 times greater than its content in other tissues. The highest content of zinc (of about 200 mg/g wet weight) is found in the peripheral zone [29], the principal region of the prostate gland that is responsible for the production and secretion of prostatic fluid rich in extraordinarily high concentrations of citrate.

Numerous *in vitro* studies indicate that zinc concentration in cancerous prostate tissue is several times lower than that in benign prostate hyperplasia (BPH) and in normal prostate tissue, however some studies reported the fact that the mean zinc concentrations in malignant and benign prostate tissues have a significant overlap [30-33], and eventually some studies present higher zinc concentration level in cancerous prostate tissues in comparison to the normal ones [34, 35].

It is clear that more elaborate studies are required, providing information on grade and stage correlation between malignancy development and zinc concentration levels in order to establish potential zinc-based diagnosis.

Table 1. Various sample preparation techniques and methods used for prostate tissues analysis

Authors	Sample preparation	Method	Remarks
Cortesi et all [36,37]	The absolute zinc content was measured……within **the entire core** ($4mm^3$) or within **four quarters along its length** ($1mm^3$ **segments**) **0.5 mm in diameter** and **15–20 mm long tissue cylinders** folded in two, were placed between two 2.5 mm Mylar foils, The entire core was irradiated and its **average** Zinc content was calculated	X-Ray Fluorescence (XRF)	
Györkey F et all [29]	Portion of tissues **were homogenized**	Atomic absorption spectrophotometry	Using AAS as an analytical technique the absorption of light of free atoms is measured. The sample is **atomised at a very high temperature (2500-3000 °C)** AAS called a "destructive technique", because **only solutions containing the investigated element** can be used. Solid samples should be accurately weighed and then dissolved, often using strong acids [38, 39]
Lahtonen R [31]	Solution of solid samples for AAS	Atomic absorption spectrophotometry	
Ogunlewe JO et all [40]	Weighed portions (500 mg) of prostatic tissue were **homogenized** in deionized water to produce a 5% homonegate. Equal volumes of this homogenate and 10% trichloracetic were **mixed** and then **centrifuged** at 2000 rpm for 20 minutes	Atomic absorption spectrophotometry	

Table 1. (Continued)

Authors	Sample preparation	Method	Remarks
Marczyńska A et all [41]	The material for analysis was **granulated** and divided into three aliquots weighing 271 + 83 mg each. **The third portion was digested with a mixture of concentrated nitric, sulphuric and perchloric acids (3:1:1) in silica crucibles.**	Flame atomic absorption spectrophotometry (Pay Unlearn)	
Zaichick VY [33]	Materials of transrectal puncture biopsy were freeze-dried and **homogenized**	INAA instrumental neutron activation analysis	
Sapota A et all [42]	Samples were **mineralized**	Flame atomic absorption spectrometry	
Vartsky D et all [43]	For XRF analysis tissue samples 1 cm in diameter and 1 to 2 mm thick	XRF The beam was collimated using a **3 mm** aperture, such that the diameter of the beam impinging on the sample was not more than about **6 mm**	Huge beam size

To build a theory and draw conclusions about the biochemical processes that take place at the cellular level, information about elemental concentrations with similar special resolution is needed. In most experiments, discussing zinc concentration levels [29, 31, 33, 36-43] in cancerous and non-cancerous prostate tissues, macroscopic samples or samples homogenized or mineralized prior to the experiments have been used for data collection. Table 1 presents various sample preparation techniques and methods used by the authors of papers [29, 31, 33, 36-43]. As can be seen from the data presented, analysis of these types of samples can provide only average values for zinc concentrations, which by no means are representative. But what is more important is that various sample treatments may influence the final results. It is rather impossible to follow very precisely processes occurring after adding different chemicals to the sample homogenates and influence of these chemicals on the final zinc concentration level.

Synchrotron Radiation X-Ray Fluorescence Emission (SRIXE)

Within the many methods suitable for detection of concentration of elements in biological tissues, the most preferential are multi-elemental highly sensitive analytical techniques like Synchrotron Radiation X-Ray Fluorescence Emission (SRIXE). In the presented work, synchrotron radiation (SR) was used to detect and locate selected trace elements in post-operative samples of prostate tissues at cellular and sub-cellular (especially in case of cancerous cells) resolution.

Synchrotron light, now used in all areas of science, is emitted by electrons accelerated in a magnetic field, providing a unique source of radiation which is continuous across the spectrum and is emitted in a collimated beam. Synchrotron sources have revolutionized many existing techniques and provided the basis for novel experimental approaches as well. The brightness of the synchrotron X-ray beam enabled the collection of data in a fraction of the time possible with conventional X-ray sources. Additionally there is a much lower energy deposition in the sample during irradiation, so the sample hardly suffers from damage. This character of SR is especially important for biological material which is a bad conductor of electrons and heat. Therefore, the SRIXE technique is a very adequate technique for quantitative analyses of trace elements in biological materials.

X-ray Fluorescence spectra provide fast and sensitive mean of detecting elemental concentrations in a tissue section, using high-intensity focused beam

- micro-beam - it is possible to locate the concentrations of selected elements within a few microns.

In this study, the zinc concentration level was determined in cancerous prostate tissues at different stages of the disease described by Gleason grade (2, 3 and 4). Additionally, healthy and hyperplastic prostate tissues were analysed. Presented research was carried out in order to understand the effects of altered homeostasis of zinc in the etiology of prostate cancer.

MATERIALS AND METHODS

Samples

All materials (prostate glands) were obtained during radical prostatectomies from 12 patients (patients' age: 56 - 72 years, mean age 61) at the Department of Urology, City Hospital im. Narutowicza, Krakow, Poland.

The sample preparation was performed in a controlled environment to eliminate or at least minimize any contamination effects. All samples were prepared in the same way, first material dedicated for analysis was frozen in liquid nitrogen in order to stop all living processes occurring in cells and hence to avoid any alteration in concentration levels of all elements. Then all specimens were cut into 14 μm slices with the use of cryomicrotome; for micro-SRIXE experiments over 100 samples were prepared, and each slice was mounted on Mylar foil and air-dried. For every experimental sample, an adjacent section (of the same thickness) was placed on microscopic glasses, and then stained with haematoxylin and eosin (H&E) dye.

An experienced histopathologist fully characterized H&E stained sections microscopically by dividing them into non-cancerous (healthy and hyperplasic prostate tissues) and cancerous groups. Additionally, the Gleason grade was assigned to areas of malignancy. For the number of patients from whom the parts of prostate glands had been obtained and classified as healthy, hyperplastic, Gleason grade 2 and 4 was equal to 2, except for the Gleason grade 4 group, where its count was equal 4. The histopathologist identifying areas of abnormal cell growth found the cancerous ones; and its appearance was compared with the Gleason scale system to determine where it falls on a scale of one to five. All samples used in this work were analysed by the same person.

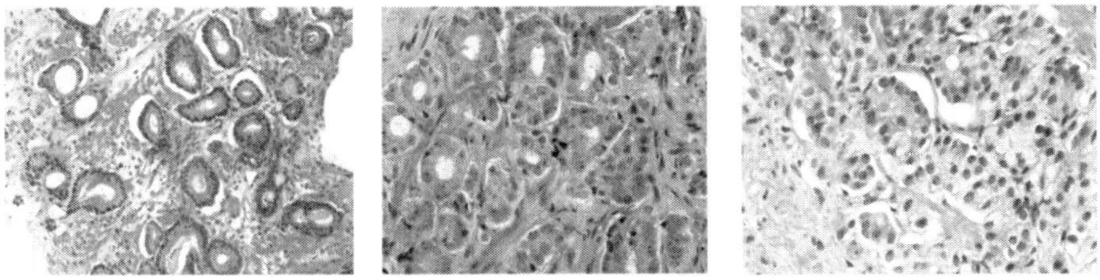

a b c

Figure 1. Haematoxylin and eosin-stained prostate cancer tissue of Gleason grades 2–4, exhibiting examples of malignant glands that were used to acquire micro-SRIXE spectra (a) Gleason grade 2, (b) Gleason grade 3, (c) Gleason grade 4.

From histopathological studies, certain areas were selected for further micro-SRIXE studies and were used as landmarks to position the micro-beam on the lesions of the unstained adjacent section. Careful specifications of selected samples were made using the Olympus BX-50 microscope. It is worth saying that any damages in sample patterns, which would be caused by Leidenfrost effect, were not observed [44]. Examples of H&E sections analysed during experiments are presented in Figure 1.

Measurements

The spectra were collected at the L-beamline at the Hasylab, DESY (Germany) using the synchrotron X-ray fluorescence microprobe. Photon flux delivered by the bending magnet at Hasylab (positron energy 4.45 GeV, storage ring current 140 mA and magnetic field 1.2182 T) was equal to 5×10^{13} photons/mrad^2/0.1%BW at 10 keV. The microprobe setup is a powerful instrument for simultaneous multi-element trace analysis of micro-samples [45-48].

The micro-SRIXE observation was conducted in air without pre-treatment and staining of samples. Using this highly brilliant synchrotron source at the Hasylab, it is routinely possible to detect the elements with atomic numbers between 13 and 92 with minimum detection limits C_{MDL} of down to 0.1 pg/μg. The white beam of a bending magnet source was monochromatized by a double-multilayer-monochromator with a bandpass $\Delta E/E \cong 2\%$. The photon flux at sample with Si(111) monochromator was equal to $9*10^9$ photons/mm^2/0.01%BW at 10 keV. The beam was focused by a polycapillary half lens (X-ray Optical Systems. Inc, focal distance 4.8 mm and transmission of 30%) and its size on the sample was approximately 15 μm (in diameter) which is of the order of a cell diameter. For this study, the energy was set to 17 keV. By using this energy, it was possible to collect fluorescence spectra with very well-determined peaks for K_α lines elements with the atomic number between 15 and 38. Fig. 2 depicts an example of spectrum collected for healthy prostate tissue.

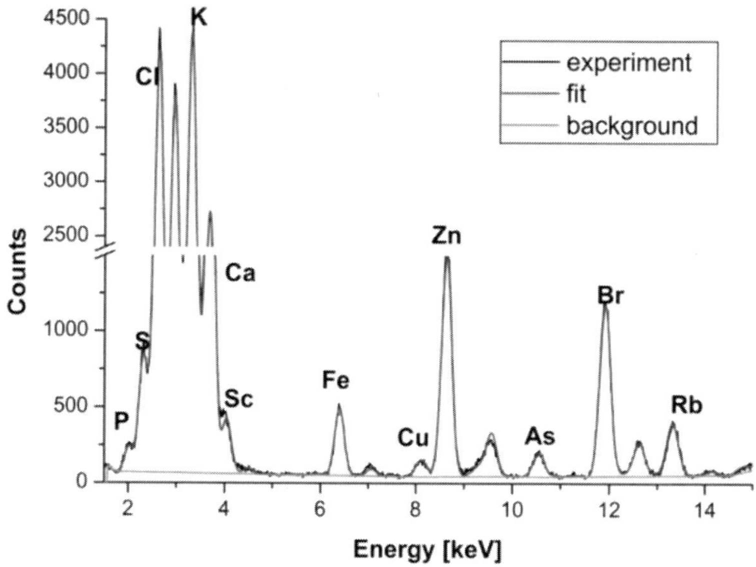

Figure 2. Micro-SRIXE spectrum collected for healthy prostate tissue.

All experiments were performed on individual samples placed on 2.5 μm Mylar foil. This foil was strong enough to support the tissue. It also could withstand focused beam. Additionally, what is most important for the analysis, micro-SRIXE spectra collected for this foil, did not contain peaks (fluorescence signal) characteristic for the elements of interest, which allowed treating the foil as a "clean" support for samples standing.

During analyses, the samples were fixed on a remote-controlled stage equipped with high precision stepping motors allowing for micrometric movements in xyz space (reproducible positioning of about 0.5 μm). The samples on the special holder were positioned at an angle of 45^0 with respect to the incoming beam, just in front of polycapillary presented in Figure 3.

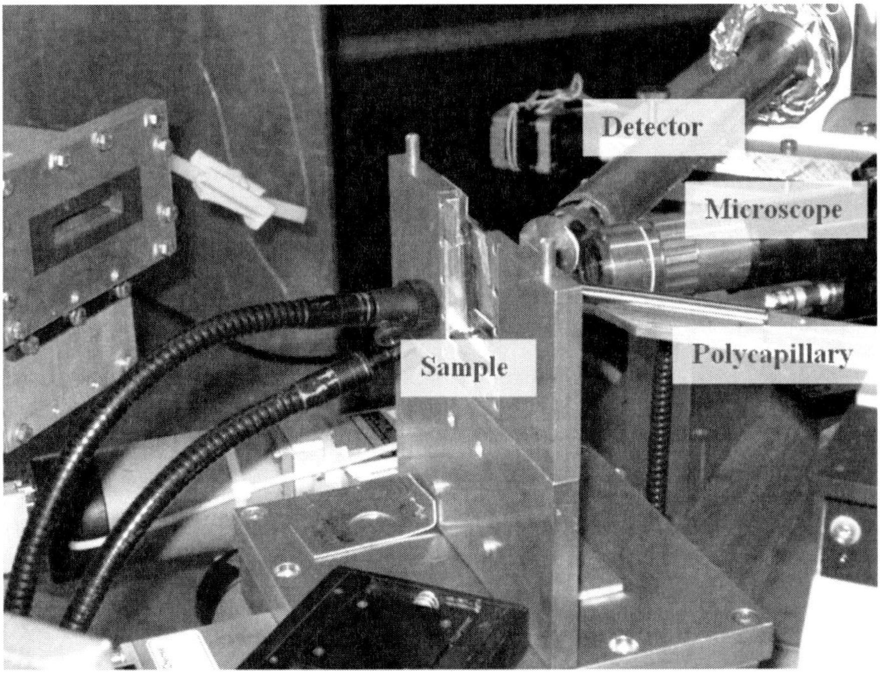

Figure 3. Photo of the experimental table showing sample holder arrangement at the L-beam line in HASYLAB.

X-ray fluorescence was detected by use of a conventional HPGe X-ray detector (with an energy resolution 120 eV at 5.89 keV) positioned at 90^0 to the incoming linear polarized X-ray beam and placed in the storage ring plane in order to minimize the intensity of the scattering X-rays. Thickness of the beryllium window of detector was 8 μm. Air path between sample and detector was set to a few millimetres. It was done after several tests to optimize the conditions of experiments. The acquisition time of the single spectrum was equal to 300 s in order to obtain reliable spectra with good signal to noise ratio. Only previously selected by histopathologist areas were analysed. On the same type of cell packaging, single points were measured with the distance around 15-20 μm between them. In case of any significant discrepancies between the spectra collected for these points, additional histopathologist's examinations were performed.

The collected spectra were deconvoluted by using AXIL (Analysis of X-Ray spectra by Iterative Least Squares regression) programme [49]. The evaluation of energy dispersive X-ray spectra by means of AXIL was achieved by fitting the measured spectra with suited mathematical functions, using a non-linear least squares strategy. Special model including information about Gaussian line shape centred around proper energy, lines positions, their relative amplitudes, distance between K_α and K_β lines, and the shape of the background etc., was created prior to the actual spectrum analysis using X-ray library available in software. In the presented analysis, the smooth filter (SF) background was chosen. SF procedure is based on the removal of rapidly varying structures in a spectrum by comparing the channel content y_i with the channel content of its neighbours. For each spectrum, the background was subtracted. Additionally all spectra were normalized to the value of the incident photon flux.

For dry mass concentrations of the elements in the analyzed samples (expressed in [μg/g]), two home-made standards were prepared by mixing pure acrylamide, *N,N*'-methylenebisacrylamide with different aqueous solutions of metal nitrates [50]. Standard MULT I composed of the mixture of nitric metals such as: Ca, V, Fe, Zn, As, Pb, Sr, Zr, Mo and Cd whereas MULT II included: K, V, Mn, Fe, Zn, Hg, Pb, Sr, Zr and Mo. Metal concentration in those standards were at the level of 10 μg/g, which was additionally confirmed by atomic absorption spectroscopy (AAS).

For calculation of the elemental level concentrations "the external standard method" was used. This method is first based on analysing the multi-elemental standards with known concentrations of elements, and then under the same experimental conditions, samples analysis is performed. Having the spectra for standard (area under the peak for selected element) so called efficiency curve is calculated in order to obtain the efficiency of detection for single element. This value strongly depends on the atomic number (Z). Taking into account also normalization parameters for the given experiment, the concentration for the selected element can be easily revealed [51].

RESULTS AND DISCUSSION

Statistical Analysis

Statistical analysis and graphic presentation of data were done in R platform (version 2.10) [52]. In order to check the normality of distribution of obtained concentrations and select the proper method for further analysis of collected data, the Shapiro-Wilk test was performed. This test is tailored to access the goodness of fit to the normal distribution and is considered a better choice than the Kolmogorov-Smirnof and Lilliefors' tests. According to the results of this test, all distribution of concentrations of selected elements are not normal (with p-value < 0.05), because several factors may contribute to the concentration range of the results, such as differences in diet, hormonal status, age, genetics, drug administration etc.

Over 300 spectra collected during experiments from different samples were taken into account in statistical analysis. The average number of spectra for each group of prostate tissues (healthy, hyperplastic, Gleason grade 2-4) was equal to 60. For zinc concentration levels, basic descriptive statistical parameters were calculated (median, minimum and maximum, skewness and curtosis values). Median values have been chosen since the concentration distributions of the elements in biological samples typically are asymmetric. Almost all distributions of zinc content within the different groups of specimens are positively skewed, what is characteristic for biological media [53].

The violin plots [54] were selected to present the results of statistical analysis. They are similar to box plots, except that they also show the probability density of the data at different values. As shown in Fig. 4, the violin plot includes a box plot with two slight modifications. Firstly, a circle replaces the median line which facilitates quick comparisons when viewing multiple groups. Secondly, outside points which are traditionally classified as outliers are not identified by individual symbols. The kernel density estimation overlaid on the box plot supplements traditional statistics by graphically showing the distributional characteristics of data.

Zinc Concentration

It was mentioned in the Introduction, that the prostate contains the highest concentration of Zn of all the soft tissues. According to works presented by Ho et al [55] and Sapota et al [42], in cases of prostate cancer, this level decreases rapidly. However our studies reported earlier [15] showed that prostate cancer tissues contain elevated concentrations of Zn when compared to normal (healthy) prostate tissues.

Based on current work, it is still possible to sustain the hypothesis mentioned earlier about the influence of Zn on prostate cancer, especially taking into account results obtained for Gleason grade 2 (GG2). The median value for Zn concentration in this group is almost 6 times higher compared to the value found for healthy prostate tissue. This highest Zn concentration level can illustrate the fact that every highly proliferating cell system is dependent on the sufficient availability of Zn [56].

For Gleason grade 3 and 4 groups (GG3 and GG4), the calculated median value for Zn is higher as compared to the healthy group, but at the same time, a gradual decrease of this value is observed with more advanced stages of prostate cancer (GG3 and GG4).

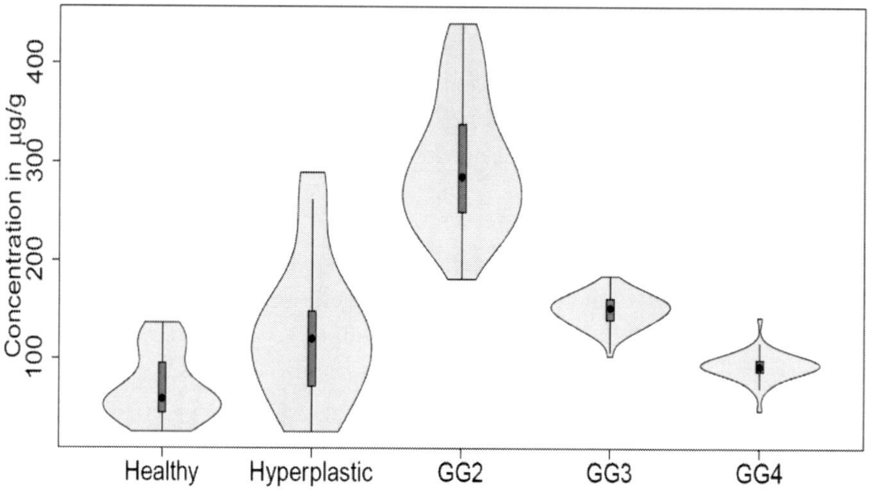

Figure 4. Overview of statistical analysis of Zn concentration in tissue specimens – violin plots for 5 categories (healthy, hyperplastic, GG2, GG3, GG4).

Distributional characteristics of data have a significant overlap for almost all pairs (shown in Tab. 2) which points to at least partially similar functions of zinc in the subsequent stages of abnormality (e.g. healthy-hyperplastic, GG2-GG3). Wide spread Zn concentration values is observed in hyperplastic and GG2 groups (Fig. 4) which may be an indication of dynamic metabolism of this element exactly in these groups._Presented here, the violin plots synergistically combine the information that the box plots usually contain and the density trace into a single display that reveal structures found within the data.

The elevated concentration of Zn possibly leads to the development of various disorders in prostate gland, higher concentration of Zn in hyperplastic prostate tissues compared to healthy ones was also observed by Sapota et al [42].

Table 2. Correlation between distributional characteristics of data calculated for various group of prostate tissue

	Healthy	Hyperplastic	Gleason grade 2	Gleason grade 3	Gleason grade 4
Healthy	X	very strong	none	strong	strong
Hyperplastic		X	strong	very strong	very strong
Gleason grade 2			X	medium	none
Gleason grade 3				X	strong
Gleason grade 4					X

Summary of Statistical Results

To validate collected results and confirm their statistical importance, Wilcoxon signed rank for the unpaired samples test and the p-values were calculated. The p-values are the products of nonparametric comparison between the groups representing the healthy and diseased tissue samples. P values lower than the cut-off point of 0.05 indicated that the two groups differed significantly with respect to the property that was used for then comparison. The obtained results indicate that in most of the cases, the median value of Zn differentiates significantly in the analysed groups ($p<0.001$). Weaker dependence is observed in distinguishing between hyperplastic - GG3 ($p<0.01$), and hyperplastic - GG4 groups ($p<0.05$).

CONCLUSION

Presented studies were undertaken to establish the variations in Zn concentrations between healthy, hyperplastic and different stages of prostate cancer tissues to understand the effects of altered homeostasis of this element in the etiology of prostate cancer. For quantitative analysis, a beam collimated to 15 μm, with the live view of the position of the beam with respect to the sample was used. There was strong identification of the tissue/cell composition that the beam was detecting. The evaluation of the grading (Gleason grade) of the chosen areas was double-checked by a histopathologist. Only areas selected earlier by the experienced histopathologist were analysed in order to obtain reliable information about zinc concentrations. Presented in this chapter is one of these techniques, in our opinion, that can give complete control over what area of tissue sample is being analysed. Only analysis performed at the cellular level can provide reliable and true information about the concentration of any element in a selected area. Fig. 5 shows a histological view of human tissue with a map of colour coded distribution of zinc concentration. One can see the correspondence between the tissue constituents and the distribution of the element concentration. The main goal of this presentation is to illustrate that, despite the fact that this sample was described as cancerous by the histopathologist, it is highly non uniform and, depending on the point chosen, different values of zinc concentration may be obtained.

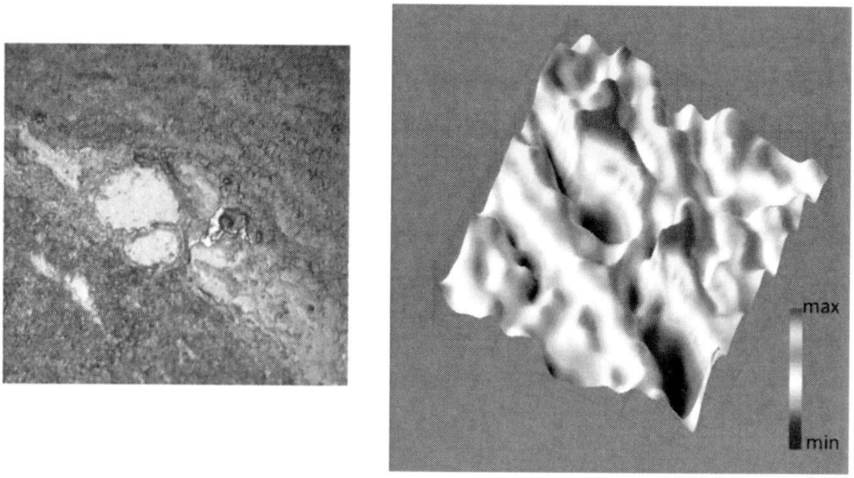

Figure 5. Histological view of human tissue (left) with the map of colour coded distribution of zinc concentration (right).

The combination of element distribution patterns of prostate tissues with an appropriate statistical technique can be used as a useful tool for distinguishing normal tissues from various stages of cancer. Taking into account the fact that the malignant process is a long-term effect of complicated biochemical processes occurring within single cells, it is necessary to present elemental concentration results separately for various cancerous stages. Many researchers [29, 31, 33, 36-43] report only results for two or three groups (healthy, cancerous, and sometimes hyperplastic).

As can be seen in Figure 4, the highest concentration of zinc is observed for samples with Gleason grade 2 (early stage of cancer development), when exactly the same data are divided into only three categories. The calculated mean values for cancerous tissues differ significantly from the values obtained for three cancerous groups – GG2, GG3 and GG4, depicted in Fig. 6. Additionally significant overlap in distribution of zinc concentration is presented for all three groups.

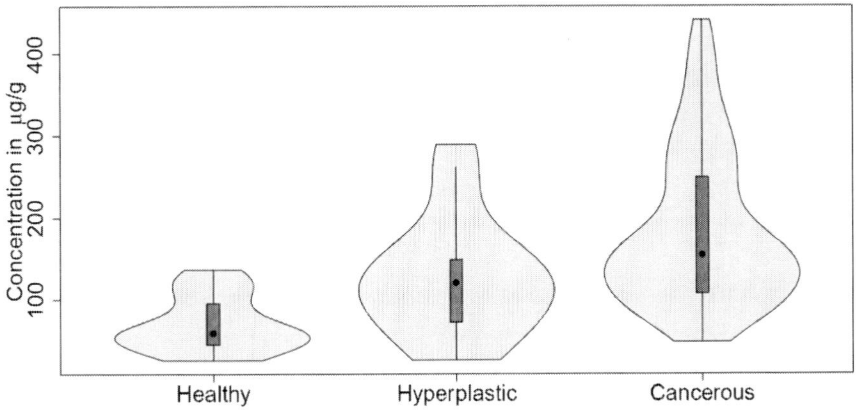

Figure 6. Violin plots presenting the distribution of the zinc concentration for another category assignment.

Another issue is the problem of finding representative samples for the healthy category. Often, tissues are collected for this purpose during *post mortem* examination. In this case it is hard to say that the patient was healthy. Even if histopathology of the prostatic gland does not reveal any cancerous changes, zinc and other elemental concentrations could be altered owing to *ante mortem* stress and *post mortem* changes.

The elevated level of Zn concentrations (median value) observed in the present work may indicate that this element is in demand in neoplastic transformation and tumour growth, which on the other hand might also be a consequence of the disease. High intraprostatic zinc concentrations may adversely affect prostate cancer risk. Zinc enhances the activity of telomerase [57], an enzyme thought to be responsible for unlimited proliferation of tumour cells and whose activity is increased in prostate cancer [58]. The prostate cancer can definitely be treated as one of the hormonally-induced cancers. Higher zinc concentrations, in comparison to healthy and hyperplastic tissues, found in this research for cancerous ones may point to increased levels of this element connected with specific types of cancers. Higher levels of zinc were also found in breast cancer [26]. Several findings that link zinc with the suppression of prostate cancer cell growth [59-61] and inhibition of prostate tumour cell invasion [62, 63] suggest that high intraprostatic zinc levels may protect against prostate carcinogenesis. Maybe observed in this work gradual decrease of zinc median value with more advanced stage of prostate cancer (GG2, GG3, GG4) is an example of reduced defensive capability of zinc in these cells.

It is widely known that normal and cancerous tissues differ biochemically and histologically. Tumours, characterized by unregulated multiplication of cells, need an increasing supply of essential nutrients including trace elements. This probably results in an increased vascularity of malignant tissues, which in turn leads to the enhancement of elemental concentrations in tumours. The malignant process that has already started can result in a lowered concentration of microelements, what can be seen as a decrease in Zn concentration (Fig. 4) for all Gleason groups. The lowering of elemental content in tumorous tissue indicates the transport changes for some elements and a number of channels of the cell membrane.

The obtained results testify the promising nature of the given study for understanding tissue degradation processes. The considerations relating to Gleason grading have called for the development of diagnostic tools that operate on the basis of cellular biochemical rather than visual tissue architectural recognition. Maybe manipulation of trace elements can provide new lines of treatment for cancers. It is clear that further studies with a larger number of tissue samples must be performed in order to consolidate the present conclusions. Moreover, the analysis and classification methodology presented in this work can be refined in the future, including other types of tissues.

As was presented here, there is still a lack of a well-established relationship between zinc and prostate cancer, and so an accurate understanding of the zinc relationship and its tumour suppressor activities is essential. To accomplish this task, an exhaustive pathology work in each area of the prostate sample is necessary, the number of cases studied needs to increase as well as reliable information about the concentration distribution of the essential elements with decent special resolution.

References

[1] Marks, S. Prostate & Cancer: A Family Guide to Diagnosis, Treatment & Survival; 3rd ed.; MA: Perseus Publishing, Cambridge, 2003;

[2] Held–Warmkessel, J. In *Cancer Nursing: Principles and Practice;* Yarbro, C. H.; Frogge, M. H.; Goodman, M.; Jones & Bartlett: Sudbury, Mass, 2005; pp. 1552–1580.

[3] Kirby, R. S.; Brawer, M. K. *Prostate Cancer;* Health Press (UK), Abington, 2004;

[4] Ramon, J.; Denis, L. J. Recent Results in Cancer Research: Prostate cancer; *Springer–Verlag,* Berlin, 2007;

[5] Gleason, D.F. In Urologic pathology: The prostate; Tannenbaum, M.; *Lea and Febringer:* Philadelphia, 1977; pp. 171–197.

[6] Smith, E. B.; Frierson, H. F.; Millis, S. E.; Boyd, J. C.; Teodorscu, D. *Cancer.* 2002, 94, 2282–2287.

[7] Lattouf, J. B.; Saad, F. *BJU INT.* 2002, 90, 694–699.

[8] Epstein, J. I.; Pizov, G.; Walsh, P. C. *Cancer.* 1993, 71, 3582–3593.

[9] Naga Raju, G. J.; Sarita, P.; Ravi Kumar, M.; Ramana Murty, G. A.; Seetharami Reddy, B.; Lakshminarayana, S.; Vijayan, V.; Rama Lakshmi, P. V.; Satyanarayana, G.; Bhuloka Reddy, S. *Nucl. Instrum. Meth.* B. 2006, 247, 361–367.

[10] Geraki, K.; Farquharson, M. J.; Bradley, D. A. *Phys. Med. Biol.* 2002, 47, 2327–2339.

[11] Geraki, K.; Farquharson, M. J.; Bradley, D. A. *Phys. Med. Biol.* 2004, 49, 99–110.

[12] Geraki, K.; Farquharson, M. J.; Bradley, D. A.; Hugtenburg, R. P. *Nucl. Instrum. Meth.* B. 2004, 213, 564–568.

[13] Guntupalli, J. N.; Padala, S.; Gummuluri, A. V.; Muktineni, R. K.; Byreddy, S. R.; Sreerama, L.; Kedarisetti, P. C.; Angalakuduru, D. P.; Satti, B. R.; Venkatathri, V.; Pullela, V. B.; Gavarasana, S. *Eur. J. Can. Prev.* 2007, 16, 108–115.

[14] Kwiatek, W. M.; Banas, A.; Banas, K.; Gajda, M.; Galka, M.; Falkenberg, G.; Cichocki, T. *J. Alloy Compd.* 2005, 401, 178–183.

[15] Kwiatek, W. M.; Banas, A.; Gajda, M.; Galka, M.; Pawlicki, B.; Falkenberg, G.; Cichocki, T. *J. Alloy Compd.* 2005, 401, 173–177.

[16] Andreini, C.; Banci, L.; Bertini, I.; Rosato, A. *J. Proteome Res.* 2006, 5, 3173–3178.

[17] Maverakis, E.; Fung, M. A.; Lynch, P. J.; Draznin, M.; Michael, D .J.; Ruben, B.; Fazel, N. J. *Am. Acad. Dermatol.* 2007, 56, 116–124.

[18] Rink, L.; Gabriel, P. *Proc. Nutr.* Soc. 2000, 59, 541–552.

[19] Ibs, K. H.; Rink, L. *J. Nutr.* 2003, 133, 1452S–1456S.

[20] Abdel–Mageed, A. B.; Oehme, F. W. *Vet. Hum. Toxicol.* 1990, 32, 34–39.

[21] Cousins, R. J. In *Present Knowledge in Nutrition;* Ziegler, E. E.; Filer, L. J.; ILSI: Washington DC, 1996; pp. 293–306.

[22] Wood, R. J.; Suter, P. M.; Russell, R. M. *Am. J. Clin. Nutr.* 1995, 62, 493–505.

[23] Shukla, V. K.; Adukia, T. K.; Singh, S. P.; Mishra, C.P.; Mishra, R. N. *J. Surg. Oncol.* 2003, 84, 31–35.

[24] Beck, F. W.; Kaplan, J.; Fine, N.; Handschu, W.; Prasad, A. S. *J. Lab. Clin. Med.* 1997, 130, 147–156.

[25] Gregoriadis, G. C.; Apostolidis, N. S.; Romanos, A. N.; Paradellis, T. P. *Cancer.* 1983, 52, 508–519.

[26] Margalioth, E. J.; Schenker, J. G.; Chevion, M. *Cancer.* 1983, 52, 868–872.

[27] Diez, M.; Arroyo, M.; Cerdan, F. J.; Munoz, M.; Martin, M. A.; Balibrea, J. L. *Oncology.* 1989, 46, 230–234.

[28] Yoshida, D.; Ikeda, Y.; Nakazawa, S. *J. Neuro–Oncol.* 1993, 16, 109–115.

[29] Gyorkey, F.; Min, K. W.; Huff, J. A.; Gyorkey, P. *Cancer.* 1967, 27, 1348–1353.

[30] Costello, L. C.; Franklin, R. B.; Liu, Y.; Kennedy, M. C. *J. Inorg. Biochem.* 2000, 78, 161–165.

[31] Lahtonen, R. *Prostate.* 1985, 6, 177–183.

[32] Habib, F. K.; Mason, M. K; Smith, P. H.; Stitch, S. R. *Br. J. Cancer.* 1979, 39, 700–704.

[33] Zaichick, V. Y.; Sviridova, T. V.; Zaichick, S. V. *Int. Urol. Nephrol.* 1997, 29, 565–574.

[34] Yaman, M.; Atici, D.; Bakirdere, S.; Akdeniz, I. *J. Med. Chem.* 2005, 48, 630-634.

[35] Banas, A.; Kwiatek, W. M.; Banas, K.; Gajda, M.; Pawlicki, B.; Cichocki, T. *J. Biol. Inorg. Chem.* 2010, 15, 1147–1155.

[36] Cortesi, M.; Fridman, E.; Volkov, A.; Shilstein, S. Sh.; Chechik, R.; Breskin, A.; Vartsky, D.; Raviv, G.; Ramon, *J. Prostate.* 2010, 70, 1701–1708.

[37] Cortesi, M.; Fridman, E.; Volkov, A.; Shilstein, S. Sh.; Chechik, R.; Breskin, A.; Vartsky, D.; Kleinman, N.; Kogan, G.; Moriel, E.; Gladysh, V.; Huszar, M.; Ramon, J.; Raviv, G. *Prostate.* 2008, 68, 994–1006.

[38] Skoog, D. A.; West, D. M.; Holler, F. J. Fundamentals of analytical chemistry; *Saunders College Publishing,* Fort Worth, 1992;

[39] Kenkel, J. Analytical chemistry for technicians; *Lewis Publishers*, Boca Raton, 1994;

[40] Ogunlewe, J. O.; Osegbe, D. N. *Cancer*. 1989, 63, 1388–1392.
[41] Marczynska, A.; Kulpa, J.; Lenko, J. *Int. Urol. Nephrol.* 1983, 15, 257–265.
[42] Sapota, A.; Darago, A.; Taczalski, J.; Kilanowicz, A. *Biometals*. 2009, 22, 1041–1049.
[43] Vartsky, D.; Shilstein, S.; Bercovich, A.; Huszar, M.; Breskin, A.; Chechik, R.; Korotinsky, S.; Malnick, S. D.; Moriel, E. *J. Urol.* 2003, 70, 2258–2262.
[44] Leidenfrost, J. G. *Int. J. Heat Mass Transfer*. 1966, 9, 1153–1166.
[45] www.hasylab.de
[46] Chwiej, J.; Janeczko, K.; Marciszko, M.; Czyzycki, M.; Rickers, K.; Setkowicz, Z. *J. Biol. Inorg. Chem.* 2010, 15, 283-289.
[47] Falkenberg, G.; Clauss, O.; Swiderski, A.; Tschentscher, T. *X-ray Spectrom.* 2001, 30, 170-173.
[48] Al-Ebraheem, A.; Farquharson, M. J.; Ryan, E. *Appl. Radiat. Isot.* 2009, 67, 470-474.
[49] Van Espen, P.; Janssens, K.; Swensters, I. *AXIL X-Ray Analysis Software Users Manual,* Benelux, 1989.
[50] Kwiatek, W. M.; Hanson, A. L.; Paluszkiewicz, C.; Galka, M.; Gajda, M.; Cichocki, T. *J. Alloy Compd.* 2004, 362, 83-87.
[51] Kwiatek, W. M. In Fizyczne metody badan w biologii, medycynie i ochronie srodowiska; Hrynkiewicz, A. Z.; Rokita, E.; *PWN,* 1999; pp. 366-406.
[52] http://www.r-project.org *R Development Core Team*, 2009.
[53] Altman, D. G. Practical Statistics for Medical Research; Chapman and Hall/CRC, London, 1991;
[54] Hintze, J. L.; Nelson, R. D. *The American Statistician*. 1998, 52, 181-184.
[55] Ho, E.; Song, Y. Current Opinion In Clinical Nutrition And Metabolic Care. 2009, 12, 640-645.
[56] Rink, L.; Gabriel, P. *Biometals*. 2001,14, 367-383.
[57] Nemoto, K.; Kondo, Y.; Himeno, S.; Suzuki, Y.; Hara, S.; Akimoto, M. *Biochem. Pharmacol.* 2000, 59, 401–405.
[58] Sommerfeld, H. J.; Meeker, A. K.; Piatyszek, M. A.; Bova, G. S.; Shay, J. W.; Coffey D. S. *Cancer Res.* 1996, 56, 218–222.
[59] Liang, J. Y.; Liu, Y. Y.; Zou, J.; Franklin, R. B.; Costello, L. C.; Feng, P. *Prostate*. 1999, 40, 200–207.
[60] Feng, P.; Liang, J. Y.; Li, T. L.; Guan, Z. X.; Zou, J.; Franklin, R. *Mol. Urol.* 2000, 4, 31–36.

[61] Iguchi, K.; Hamatake, M.; Ishida, R.; Usami, Y.; Adachi, T., Yamamoto, H. *Eur. J. Biochem.* 1998, 253, 766–770.

[62] Ishii, K.; Usui, S.; Sugimura, Y.; Yoshida, S.; Hioki, T.; Tatematsu, M. *Int. J. Cancer.* 2001, 92, 49–54.

[63] Ishii, K.; Usui, S.; Sugimura, Y.; Yamamoto, H.; Yoshikawa, K.; Hirano, K. *Biol. Pharm. Bull.* 2001, 24, 226–230.

In: Zinc: Characteristics, Uses and Benefits ISBN: 978-1-61942-523-1
Editor: Masayoshi Yamaguchi

Chapter 4

ZINC AND ATHEROSCLEROSIS

Bin Bao*

Department of Pathology, Karmanos Cancer Institute, School of Medicine, Wayne State University, Detroit, MI, US

ABSTRACT

Atherosclerosis is a chronic inflammatory process, triggered by the presence of lipids in the vascular wall. The disease is associated with complex interactions among inflammatory cells, vascular elements, and lipoproteins, along with the expression of selected adhesion molecules and inflammatory cytokines. The strategy for protection against atherosclerosis has been a subject of intense research over the past decades. The important role of zinc has been recognized in various biochemical and physiological functions. Early evidence indicates that zinc deficiency can cause growth retardation, delayed sex maturation, depressed immune response, and abnormal cognitive functions. Nutritional zinc deficiency is common in the developing countries. 30-40% of the elderly populations are estimated to have mild to modest zinc deficiency in the USA. Conditioned zinc deficiency is also present in many chronic diseases such as rheumatoid arthritis, diabetes, and cancers. The evidence from experimental, clinical, and epidemiological studies suggests that zinc has an atheroprotective function, potentially due to its anti-inflammatory and anti-oxidative stress effects, and other functions. However, the exact mechanism of zinc action in atherosclerosis is not

* E-mail: baob@karmanos.org, bbao@med.wayne.edu

fully understood. A large number of studies demonstrate that zinc deficiency increases the levels of inflammatory cytokines and oxidative stress, and induces apoptosis and endothelial cell dysfunction. Zinc supplementation reverses these adverse effects. Our recent *cell culture* studies have also demonstrated that zinc decreases the productions of oxidative stress and inflammatory cytokines, vascular cell adhesion molecule-1 (VCAM-1), as well as oxLDL-, and/or LPS-induced NF-κB activation in human monocyte-like and vascular endothelial cells. Our normal human subject study shows that zinc supplementation decreases the productions of oxidative stress biomarkers, TNF-α and IL-1β mRNAs, as well as TNF-α-induced NF-κB activation in isolated peripheral blood mononuclear cells (PMNC), compared to the placebo group. Normal elderly subjects show decreased levels of plasma and lymphocyte zinc, and increased generation of inflammatory cytokines and adhesion molecules in isolated PMNC cells, compared to normal young adult. Six months of zinc supplementation in normal elderly subjects increased plasma zinc and anti-oxidant capacity, and decreased plasma levels of C-reactive protein (CRP), IL-6, VCAM-1, MCP-1, and oxidative stress biomarkers, compared to the placebo-supplemented elderly subjects; this is consistent with the decreased incidence of infection. The data confirmed that zinc decreases the productions of inflammatory cytokines, oxidative stress, and atherosclerosis-associated biomarkers, suggesting that zinc may play a protective role in atherosclerosis through inhibition of inflammation and oxidative stress. In this chapter, we will discuss, in-depth, the molecular mechanisms of zinc action on the development and progression of atherosclerosis.

Introduction

Atherosclerosis is the most common type of cardiovascular disease (CVD), which is the leading cause of death in USA [1]. It is a chronic inflammatory process, triggered by the presence of lipids such as oxidized LDL in the vascular wall, and is highly associated with a complex interaction among inflammatory cells, vascular elements, and lipoproteins along with the expression of several adhesion molecules and inflammatory cytokines [2,3]. The classical risk factors of atherosclerosis are age, sex, smoking, hypertension, diabetes, and obesity [4-6]. However, about 40%-50% of atherosclerotic patients over 50 years old have other risk factors such as inflammation, oxidative stress, or endothelial dysfunction, which are independent of the classical risk factors [1,7,8]. The strategy for protection against atherosclerosis has been a subject of intense research over the past decades. However, the prevention and early management of atherosclerosis is still lacking due to the lack of clear understanding of the cause of this process. The essentiality of zinc has been recognized in humans over many decades. A large number of evidence has indicated that zinc plays a critical role in a wide variety of biochemical and physiological functions [9-11]. Nutritional deficiency of zinc is common in the developing countries and conditioned deficiency of zinc is present in many chronic diseases such as rheumatoid arthritis, diabetes, obesity, and cancers, which are highly associated with chronic inflammation and oxidative stress [9,11-14]. Accumulating evidence show that zinc deficiency increases the level of inflammatory cytokines and oxidative stress, and induces apoptosis and endothelial cell dysfunction. Furthermore, a number of experimental studies indicate that low level of zinc in the body is associated with the development of atherosclerosis, suggesting that zinc may have atheroprotective functions. However, the mechanism of zinc action in atherosclerosis has not been well understood. One of the detailed mechanisms of zinc's atheroprotective properties is that zinc decreases the production of atherosclerosis-related molecules such as inflammatory cytokines, oxidative stress, and endothelial cell adhesion molecules via inhibition of NF-κB activation.

ZINC: PHYSIOLOGY

Zinc is an essential trace mineral element for mammals and plays important roles in various biochemical and physiological functions [9,11,13,14]. Zinc is well known to participate in activation of approximately 300 enzymes involved in DNA synthesis, cell division, and protein synthesis, and other metabolisms [9,11,13,14]. Zinc is also required to stabilize three dimensional structures of over 2,000 transcription factors, such as "the zinc-finger" proteins. These transcription factors are involved in the gene expression of various growth factors, steroid receptors, and immune response mediators [9,11,13-15].

Clinical observation, experimental and animal studies indicate that zinc also plays important roles in immune functions. Zinc regulates several functions of lymphocytes, such as mitogenesis, antibody synthesis, the activation of T-cells and natural killer (NK) cells, and more specifically cellular immunity [16-18]. Accumulating evidence show that zinc deficiency impairs cellular immune functions. Dysfunction of cellular immunity due to zinc deficiency in humans induces frequent severe fungi, virus, and bacterial infections, thymic atrophy, anergy, reduced serum lymphocyte proliferative response to mitogens, a selective decrease in T helper cells and reduced thymic hormone (thymulin) activity. Zinc supplementation reverses all of these manifestations [19, 20], which suggest that zinc is an important immune response mediator. Recent studies show zinc has anti-inflammatory, anti-oxidant, and anti-apoptotic effects [12, 21].

Pathogenesis of Atherosclerosis

Atherosclerosis is a slowly progressing chronic inflammatory disease characterized by focal arterial lesions that can ultimately occlude the entire blood vessel and lead to sudden death. This chronic inflammation involves a combined innate and adaptive immune response to modified lipids deposited focally within the artery wall [2,3,22]. Infiltrating monocytes differentiate into macrophages within lesions and play critical roles in both the innate and adaptive immune responses, whereas resident vascular cells and infiltrating T cells mediate the adaptive response [23]. A large number of growth factors, cytokines, and chemokines released from endothelial cells, smooth muscle, macrophages, and lymphocytes are involved in this chronic inflammatory and fibroproliferative process [2,3,22]. Excessive inflammatory and immune responses driven by inflammatory cytokines appear to be responsible for most cardiovascular events associated with atherosclerosis. Large numbers of evidence show that many inflammatory and oxidative stress biomarkers are associated with the CVD patients or patients with high risk for CVD [24-30]. High sensitive C-reactive protein (hsCRP), an inflammatory marker, has been recommended as an independent risk factor by CDC [31-33]. Therefore, atherosclerosis is strongly associated with inflammation.

ROS, Lipid Peroxidation, and Atherosis

Reactive oxygen species (ROS) are produced as intermediates in reduction-oxidation (redox) processes, leading from oxygen to water [34, 35]. The univalent reduction of oxygen, in the presence of a free electron (e), yields $^{\cdot}O_2^-$ (superoxide), H_2O_2 (hydrogen peroxide) and $^{\cdot}OH$ (hydroxyl radical). Superoxide has an unpaired electron, which imparts high reactivity and renders it unstable and short-lived [36, 37]. ROS is considered as second messengers and have been implicated as important signaling molecules [35]. Similar to second messengers, production of ROS is tightly regulated by extracellular stimuli such as hypoxia and inflammatory cytokines. Redox signaling involves at least one reaction in which oxidation of a signaling molecule by a ROS occurs and which is reversible [38]. Physiologic generation of ROS has been implicated in a variety of biological responses from transcriptional activation to cell proliferation. Under pathological conditions, a disequilibrium between ROS generation and antioxidant protection results in increased bioavailability of ROS leading to a state of oxidative stress [39, 40]. The pathogenic outcome of oxidative stress is oxidative damage of cells or tissues [37], a major cause of vascular injury in cardiovascular disease [41].

Accumulating evidence indicate that production of ROS and activation of redox-dependent signaling cascades are critically involved in atherosclerosis [35,39,41]. All types of vascular cells, including endothelial cells, smooth muscle cells, adventitial fibroblasts, and resident macrophages, produce ROS [42,43]. Of particular importance in the vasculature are superoxide ($^{\cdot}O_2^{-}$) and hydrogen peroxide (H_2O_2), since these ROS act as inter- and intra-cellular signaling molecules. The major source of ROS in the vascular wall is non-phagocytic NADPH oxidase, which is regulated by vasoactive agents (angiotensin II, thrombin, and serotonin), cytokines (IL-1β and TNF-α), growth factors (IGF-1 and EGF) and mechanical forces [35,44].

Atherogenesis is considered to be a disease involving inflammatory cascades initiated by a diverse group of stimuli. The most studied and by the far the most important stimulus is lipid [45]. Extracellular accumulation of lipids occurs early in response to increased plasma lipoprotein levels in the body [45,46]. LDL is modified rapidly in the subendothelial space into minimally modified LDL (mmLDL) and then eventually into oxidized LDL (oxLDL) by ROS. The *in vitro* studies show that oxLDL induces the expression of inflammatory cytokines such as monocyte chemotactic protein-1 (MCP-1), TNF-α in smooth muscle cells (SMCs) and endothelial cells (ECs) [47,48], and itself is chemotactic for monocytes and T lymphocytes [49]. oxLDL also induces monocyte nuclear factor (NF)-κB-mediated inflammatory gene transcription, including expression of inflammatory cytokines and adhesion molecules. Cellular oxidative stress is coupled to the modification of lipids by ROS, which are generated in cellular metabolism. ROS itself can initiate NF-κB-mediated transcriptional activation of inflammatory responsive genes [50], thereby, potentially acting as independent triggers of atherosclerosis.

Zinc has been shown to function as an anti-oxidant in the body [51-54]. The evidence from cell culture, animal models, and humans show that zinc increases the activities of GSH, catalase and SOD, and decreases inducible NO synthase (iNOS) activity and lipid peroxidation product level [55,56]. Zinc-inducing metallothionein (MT) proteins also can act as an anti-oxidant against ROS [57]. Our published data from cell culture models and human subjects demonstrate that zinc decreases NO production, lipid peroxidation products, iNOS, TNF-α-, oxLDL-, or LPS-induced NF-κB activation in HL-60 (human promyelocytic leukemia cells), THP-1 (human monocytic leukemia cells), HAEC (human aortic endothelial cells), or human isolated PMNC cells [12,58,59], suggesting that zinc may reduce oxidative stress in the development and progression of atherosclerosis.

Inflammatory Cytokines, Adhesion Molecules, Atherosclerosis, and Zinc

Inflammatory cytokines and endothelial adhesion molecules play a critical role in the development and progression of atherosclerosis [60]. Numerous studies show that inflammatory cytokines such as TNF-α, IL-1β, IL-6, IL-8, and MCP-1, as well as endothelial adhesion molecules have been identified in the atherosclerotic lesion from animal models and CVD patients [24,26-28,31,33]. The inflammatory cytokines and endothelial adhesion molecules are involved in atherosclerosis in every stage, from the initiation of atherosclerosis to atheromous rupture. These cytokines and molecules are either absent or expressed at a basal level in the resting endothelial cells, leukocytes, and other type of cells, and can be up-regulated by NF-κB activation in response to many environmental stimuli such as TNF-α, IL-1β, PMA, LPS, or ROS [45].

It has been known that atherosclerosis is associated with the overproduction of inflammatory cytokines [61,62]. Some inflammatory cytokines such as TNF-α, IL-1β, IL-6, IL-8, and MCP-1 have been used for the inflammatory biomarkers of atherosclerosis [25,26,29,31-33]. A large number of evidence show that levels of soluble adhesion molecules are correlated with various cardiovascular risk factors such as smoking [63,64], hypertension [65,66], low HDL cholesterol [67] and hypercholesterolemia and/or hypertriglyceridemia [68,69]. The correlations of intercellular adhesion molecule-1 (ICAM-1) and VCAM-1 with lipids are already present at very young age [70]. Levels of soluble adhesion molecules, especially ICAM-1 and

P-selectin, were reported to be lowered by statin therapy [71,72]. Of special interest is the apparent lowering effect of hormone replacement therapy on soluble adhesion molecules in healthy post-menopausal women as well as post-menopausal women with CVD [73-75]. The mechanisms by which these molecules are down-regulated after activation are important since it is necessary to clear them from the cell surface after activation in order to limit the inflammatory process. Over-expression or maintenance of these molecules may have critical consequences on inflammation and on the integrity of the vascular wall, which contribute to atherosclerosis.

Zinc has been identified to be immune mediators against infection and inflammation [9,11,13-15]. It is proposed that zinc has anti-inflammatory properties by down-regulation of inflammatory cytokines and adhesion molecules via inhibition of NF-κB activation, a major transcription factor for inflammatory and immune response [76,77]. The evidence show that dietary zinc-depletion in normal human volunteers increased the generation of IL-1β in isolated human PMNC. Zinc repletion corrected the increased generation of this cytokine [78]. The mechanism of zinc down-regulation of inflammatory cytokines has been not well understood. According to our previous studies, zinc decreased the levels of TNF-α, IL-1β, MCP-1, VCAM-1 cytokines and mRNAs, as well as oxLDL-, PMA-, or LPS-induced NF-κB activation in HL-60, THP-1, and human vascular endothelial cells [58,59]. Our early studies in normal volunteers show that zinc supplementation decreased the levels of TNF-α and IL-1β mRNAs, as well as TNF-α-induced NF-κB activation in isolated PMNC, compared to the placebo group [12]. Normal elderly subjects show decreased levels of plasma zinc and lymphocyte zinc, and increased levels of TNF-α, IL-1, VCAM-1, ICAM-1 generated in isolated PMNC cells, compared to normal young adults. Six months of zinc supplementation to normal elderly subjects increased plasma zinc, and decreased plasma levels of hsCRP, VCAM-1, IL-6, secretary phospholipase A2 (sPLA), E-selectin, and MCP-1, compared to the placebo group [58]. These data confirmed that zinc decreases the atherosclerosis-related inflammatory markers, suggesting that zinc may have atheroprotective effects.

NF-κB, A20, Zinc, and Atherosclerosis

NF-κB is one of the major immune response transcription factors, involved in the initiation and development of atherosclerosis [50]. This factor is activated by many environmental stimuli including inflammatory cytokines

(IL-1β and TNF-α), LPS, protein kinase C activators (phorbol esters and platelet-activating factor), ROS, ultraviolet light and ionizing radiation [76,77,79]. NF-κB activation regulates the expression of numerous genes such as inflammatory cytokines (IL-1β, TNF-α, granulocyte macrophage colony stimulating factor, and IL-2), chemokines (IL-8 and MCP-1), enzymes (iNOS and inducible cyclooxygenase), adhesion molecules (ICAM-1, VCAM-1, and E-selectin), receptors (IL-2 receptor-α), and A20 (also known as TNF-α induced protein 3). TNF-α and IL-1β not only activate NF-κB but they themselves are induced by NF-κB activation [80]. This may lead to a positive feedback-loop amplification, with chronic activation of NF-κB in certain cells. NF-κB, rather than acting alone, co-operates with other transcription factors, such as AP-1 and C/EBP (NF-κB-bZip interactions), to amplify gene expression. This may explain the various range of genes expressed in different types of cells [81].

NF-κB is normally held in the cytoplasm in an inactive form by binding to an inhibitory protein of NF-κB (IκB), which has several subunits such as IκB-α, IκB-β, IκB-γ [76,79,80]. The state of NF-κB activation is determined by binding to IκB. Cytosolic NF-κB only becomes active and translocates to the nucleus when IκB has dissociated from the NF-κB heterodimer p50/p65. Many stimuli activate NF-κB via phosphorylation of IκB proteins through the action of specific kinases such as IκB kinase (IKK). Phosphorylation of IκB results in the attachment of ubiquitin residues and subsequent degradation by the multifunctional proteolytic enzyme proteasome [76,79,80]. Dissociation of IκB from NF-κB results in their rapid translocation to the nucleus for activation of NF-κB target genes

Zinc plays an important role in NF-κB activation. However, the regulation of NF-κB activation by zinc appears to be cell specific. Several studies show that zinc is required for NF-κB DNA binding in either purified proteins, recombinant NF-κB p50, or Th_0 cell lines [82-84]. However, accumulating evidence show that zinc decreases LPS-, ROS-, or TNF-α-induced NF-κB activation in endothelial cells, cancer cells, pancreatic cancer cell, and PMNC cells [85-88]. Therefore, there are other inhibitors of NF-κB activation that are different from IκB protein. One such inhibitor of NF-κB activation is A20, a zinc finger-transactivating factor that plays an important role in reducing IL-1β- and TNF-α-induced NF-κB activation [89-91]. A20 is expressed by regulation of NF-κB activation in various types of cells in response to a number of stimuli such as TNF-α, IL-1β, LPS, PMA, ROS as well as other stimuli. A20 was originally reported to protect cells from TNF-α-induced cytotoxicity by inhibiting the activation of NF-κB, which leads to decreased IL-1β and TNF-α signaling as has been demonstrated in endothelial cells [91]. It has been reported that A20 inhibits NF-κB signaling by TNF-α and IL-1β via TNF-receptor associated factor (TRAF) pathways in endothelial cells [92]. Recently, A20 has also been found to contain 2 ubiquitin-editing domains for de-ubiquitinating proteins to inhibit NF-κB activation [93,94].

The mechanism of zinc in regulation of the gene expression of inflammatory cytokines, adhesion molecules, and oxidative stress mediators has not been completely elucidated. In our human subject study, we showed that in PMNC, the TNF-α induced binding of NF-κB to DNA was decreased in the zinc supplemented subjects compared to that in the placebo group, thus demonstrating the protective effect of zinc on TNF-α-induced NF-κB activation in mononuclear cells [12]. We also observed that the expression of genes encoding TNF-α and IL-1β (inflammatory cytokines known to generate ROS) were decreased in zinc-supplemented volunteers versus that of the placebo-supplemented group [12]. In our cell model study we show that the levels of A20 protein and mRNA are increased in zinc supplemented HL-60 [12]. Recently, we demonstrate that zinc increases the expression of A20 protein, and decreases NF-κB activation in HL-60, THP-1, HAEC cells after oxLDL stimulation, compared to the zinc deficient condition [58]. Moreover, zinc increases A20 and A20-TRAF1 complex, and decreases the production of inflammatory cytokines and oxidative stress in HL-60, TPH-1, and vascular endothelial cells, compared to zinc sufficient cells [12,58,59]. Silencing of A20 by its anti-sense mRNA increased TNF-α and IL-1β production in zinc-sufficient cells [59], suggesting that zinc down-regulates the production of inflammatory cytokines via A20 pathway, which is due to deregulation of NF-κB activity in the development of atherosclerosis. Thus, the effect of zinc on NF-κB activation is different in T helper 1 (Th_1) cells where zinc activates NF-κB, whereas in monocytes/macrophages, vascular endothelial cells, and cancer cells, zinc decreases NF-κB activation via TRAF pathway by A20. Therefore, zinc-dependent A20 may play a protective role through down-regulation of NF-κB activation in vascular endothelial cells.

PPAR-α, -γ, ZINC, AND ATHEROSCLEROSIS

The peroxisome proliferator-activated receptor (PPAR) α and γ of nuclear receptors, the mediators for lipoprotein metabolism, inflammation, and glucose homeostasis, have been shown to play an important protective role in the development and progression of atherosclerosis [95]. PPAR-α and γ are extensively distributed in many tissues, and abundantly present in the vascular wall where it may exert anti-inflammatory and anti-atherogenic effects [96,97]. Recently, PPARs have been shown to have atheroprotective functions [98-101]. PPAR agonists have been utilized for treatment of CVD patients or the patients with high risk for CVD such as diabetes to prevent cardiovascular events [102]. The mechanism of PPAR-α and γ against atherosclerosis is not fully understood. Accumulating evidence shows that one of the mechanisms is due to their anti-inflammatory effect by down-regulation of atherosclerosis-related NF-κB activation via negative cross-talk in nuclear DNA binding level [103]. The evidence also shows that the activation of PPAR-α, and γ leading to the down-regulation of inflammatory cytokines and endothelial adhesion molecules in endothelial cells is considered to be zinc dependent [104].

Reiterer and colleagues [105] reported that after the zinc depletion induced by N,N,N'N'-tetrakis (2-pyridylmethyl) ethylenediamine (TPEN, a zinc specific chelator) in the porcine pulmonary arterial endothelial cells, the PPAR-α or PPAR-γ agonist-pretreated cells lost the potency to down-regulate TNF-α-induced NF-κB and AP-1 binding activities, as well as VCAM-1 and IL-6 mRNA levels. However, if zinc was added back to the zinc depleted endothelial cells, all PPAR agonists significantly down-regulated these inflammatory parameters [105]. Another study demonstrates that zinc deficiency induces vascular inflammatory parameters associated with NF-κB and PPAR signaling pathways [106]. In our recent report, HAEC cells incubated in 1 μM zinc medium (zinc deficient level) showed a decrease in the expression of PPAR-α protein after oxLDL stimulation, compared to those incubated in 15 μM zinc medium (physiological level) [58], suggesting that zinc increases the expression of PPAR-α, which may contribute to down-regulation of inflammatory cytokines and adhesion molecules in the development and progression of atherosclerosis.

Apoptosis and Zinc in Atherosclerosis

Apoptosis is a process of programmed cell death, which involves the activation of a cellular intrinsic suicide program, and is known to play a major role during development in homeostasis and disease processes [107,108]. Accumulating evidence show that apoptosis is involved in the regulation of aortic intimal thickening during atherosclerosis [109], and apoptotic cell death is common in atherosclerotic plaques [110-112].

Zinc is a potent inhibitor of apoptosis [113] as zinc deficiency can induce apoptosis [15]. The molecular mechanisms by which zinc affects apoptosis are not well understood and may occur at multiple levels. Thus, zinc may have anti-atherogenic effect by interfering with signaling pathways involved in apoptosis. For example, the intracellular pathways leading to apoptotic cell death can be modulated by selective manipulation of intracellular zinc in intact cells [114]. The anti-apoptotic effect of zinc has been attributed to its inhibition of calcium- and magnesium-dependent endonucleases, a terminal step and hallmark of apoptosis [115]. Pretreatment of cells with zinc prevented apoptosis in response to various agents including lipids and TNF-α [116]. It has been reported that the protective mechanisms of zinc partly appear to be via inhibition of the proteases that catalyze the conversion of the precursor of caspase-3 to active caspase-3, an apoptosis inducing protease [117,118]. Another role of zinc in inhibition of apoptosis is that zinc down-regulates the expression of TNF-α, an apoptotic inducer [119]. A20 has been shown to have anti-apoptotic effect on TNF-α-induced apoptosis [92,120]. In our unpublished data, zinc increased the gene expression of BCL2, an anti-apoptotic mediator, in HL-60 cells. Therefore, we hypothesize that zinc increases BCL2, resulting in the down-regulation of cellular apoptosis, potentially via A20, a zinc finger protein. Further investigation is required to clarify this mechanism.

Vascular Endothelial Cell Integrity, Atherosclerosis, and Zinc

The integrity of endothelium is required for the maintenance of vascular endothelial cell function. Dysfunction of vascular endothelial cells is strongly associated with the development and progression of atherosclerosis. It is known that zinc is involved in the stabilization of DNA and protein structures. Accumulating evidence suggest that zinc plays an important role in maintaining vascular endothelial cell integrity during stressful conditions such as an inflammatory response in atherosclerosis. Low dietary zinc intake has been found to be associated with carotid intima-media thickness in the human adult subject free of clinic CVD [121], suggesting that zinc may exert an atheroprotective effect by maintaining vascular endothelial cell function. However, the detailed mechanism of zinc action on vascular endothelial cell integrity is not fully understood. Several experimental studies show that zinc deficiency increases apoptosis and disruption of vascular endothelial barrier function induced by lipids and inflammatory cytokines [21,122-126]. A recent study demonstrates that the over-expression of A20, a zinc-dependent anti-inflammatory protein, inhibits monocyte adhesion, and protects vascular endothelial cells from high glucose-induced injury, potentially through inactivation of NF-κB [127]. Therefore, the critical role of zinc in the maintenance of vascular endothelial cell integrity may be due to its anti-apoptotic and anti-inflammatory functions through A20 signaling pathway.

Lipid, Atherosclerosis, and Zinc

Dyslipidemia is one of the major risk factors in atherosclerosis. Low HDL, high LDL, especially oxidized LDL are the detrimental factor for atherosclerosis [128,129]. The role of zinc in lipid metabolism has not been defined. The results of zinc on plasma lipid profile are controversial or inconsistent in animal and human studies. Reiterer and colleagues [130] report that mice fed with a zinc-deficient diet (0 μM/g zinc) had significantly increased concentrations of cholesterol, triacylglyceride in the VLDL and HDL fraction compared to those fed zinc supplemented diet (1.529 μM/g zinc), which is consistent with other reports in rabbit models [131,132]. Some investigators found an increased concentration of cholesterol, phospholipids, VLDL and intermediate density lipoproteins in zinc-deficient rats [55]. Others have reported no effect of zinc on lipid metabolism [133,134]. This inconsistency may be due to the different species of animals used and different doses of zinc.

The atherosclerotic aortas are shown to have decreased level of zinc, compared to normal aortas [128]. One early clinical study reported the beneficial effect of high dose of zinc treatment in angina pectoris. Short term of high dose of zinc treatment (300 mg zinc daily for 3-4 weeks) to the atherosclerotic patients decreased the incidence of angina pectoris, even though increased levels of plasma LDL cholesterol was observed in the atherosclerotic patients with a high dose of zinc treatment [128]. A large number of recent clinical observations demonstrate that low level of serum zinc, increased urine zinc loss, and decreased ratio of serum zinc/copper are associated with the development of atherosclerosis [135-139].

An early epidemiological study shows that dietary zinc intake is related to CVD. In the Iowa Women's Health Study [140], 34492 postmenopausal women were recruited for the completion of food-frequency questionnaires, and followed for CVD mortality over 15 years. The results showed that the women with higher intake of zinc had lower risk of CVD mortality, and dietary zinc intake had an inverse association with CVD mortality. One recent epidemiological study confirms that dietary zinc intake is inversely related to the development of atherosclerosis [121]. These data suggest that zinc plays an important role in prevention of atherosclerosis.

CONCLUSION

In summary, inflammation, oxidative stress, and apoptosis play a critical role in the initiation and progression of atherosclerosis. The production of inflammatory cytokines, adhesion molecules, and ROS products are regulated by atherosclerosis-related NF-κB activation, in response to detrimental stimuli. As discussed above in this chapter, zinc may have atheroprotective property by the mechanisms that involve in multiple signaling transductions of multiple molecules/mediators. Since zinc is non-toxic, non-mutagenic, and cost-effective, it may be advantageous to use zinc for prevention of atherosclerosis in humans. However, further mechanistic and well-designed clinical trials are needed in order to fully appreciate the health benefits of zinc for athero-sclerosis.

ACKNOWLEDGMENTS

We thank Ms. Amanda C. Sehmer and Ms. Ginny W. Bao for editing the manuscript.

REFERENCES

[1] C.H. Hennekens, *Circulation* 97, 1095 (1998).

[2] P. Libby, *Nature* 420, 868 (2002).

[3] R. Ross, *N. Engl. J. Med.* 340, 115 (1999).

[4] G.S. Berenson, S.R. Srinivasan, W. Bao, W.P. Newman, III, R.E.T racy, and W.A. Wattigney, *N. Engl. J. Med.* 338, 1650 (1998).

[5] S.M. Grundy, G.J. Balady, M.H. Criqui, G. Fletcher, P. Greenland, L.F. Hiratzka, N. Houston-Miller, P. Kris-Etherton, H.M. Krumholz, J. LaRosa, I.S. Ockene, T.A. Pearson, J. Reed, R. Washington, and S.C. Smith, Jr., *Circulation* 97, 1876 (1998).

[6] U.N. Khot, M.B. Khot, C.T. Bajzer, S.K. Sapp, E.M. Ohman, S.J. Brener, S.G. Ellis, A.M. Lincoff, and E.J. Topol, *JAMA* 290, 898 (2003).

[7] E. Braunwald, *N. Engl. J. Med.* 337, 1360 (1997).

[8] L.G. Futterman and L. Lemberg, *Am. J. Crit Care* 7, 240 (1998).

[9] A.S. Prasad, *Nutrition* 11, 93 (1995).

[10] A.S. Prasad, *Mol. Cell Biochem.* 188, 63 (1998).

[11] A.S. Prasad, *Exp. Gerontol.* 43, 370 (2008).

[12] A.S. Prasad, B. Bao, F.W. Beck, O. Kucuk, and F.H. Sarkar, *Free Radic. Biol. Med.* 37, 1182 (2004).

[13] A.S. Prasad, Annu. Rev. Pharmacol. Toxicol. 19, 393 (1979).

[14] A.S. Prasad, *J. Am. Coll. Nutr.* 7, 377 (1988).

[15] A.H. Shankar and A.S. Prasad, *Am. J. Clin. Nutr.* 68, 447S (1998).

[16] L.D. Antoniou, R.J. Shalhoub, and G.P. Schechter, *Am. J. Clin. Nutr.* 34, 1912 (1981).

[17] P.J. Fraker, M.E. Gershwin, R.A. Good, and A. Prasad, *Fed. Proc.* 45, 1474 (1986).

[18] A.S. Prasad, S. Meftah, J. Abdallah, J. Kaplan, G.J. Brewer, J.F. Bach, and M. Dardenne, *J. Clin. Invest* 82, 1202 (1988).

[19] K.H. Falchuk, *Mol. Cell Biochem.* 188, 41 (1998).

[20] N. Wellinghausen and L. Rink, *J. Leukoc. Biol.* 64, 571 (1998).

[21] B. Hennig, P. Meerarani, M. Toborek, and C.J. McClain, *J. Am. Coll. Nutr.* 18, 152 (1999).

[22] R. Ross, *Nature* 362, 801 (1993).

[23] G.K. Hansson, *N. Engl. J. Med.* 352, 1685 (2005).

[24] J.A. de Lemos, C.H. Hennekens, and P.M. Ridker, *J. Am. Coll. Cardiol.* 36, 423 (2000).

[25] A. Deswal, N.J. Petersen, A.M. Feldman, J.B. Young, B.G. White, and D.L. Mann, *Circulation* 103, 2055 (2001).

[26] A.E. Hak, H.A. Pols, C.D. Stehouwer, J. Meijer, A.J. Kiliaan, A. Hofman, M.M. Breteler, and J.C. Witteman, *J. Clin. Endocrinol. Metab* 86, 4398 (2001).
[27] S.J. Hwang, C.M. Ballantyne, A.R. Sharrett, L.C. Smith, C.E. Davis, A.M. Gotto, Jr., and E. Boerwinkle, *Circulation* 96, 4219 (1997).
[28] L. Lind, *Atherosclerosis* 169, 203 (2003).
[29] T.A. Pearson, G.A. Mensah, R.W. Alexander, J.L Anderson, R.O. Cannon, III, M. Criqui, Y.Y. Fadl, S.P. Fortmann, Y. Hong, G.L Myers, N. Rifai, S.C. Smith, Jr., K. Taubert, R.P. Tracy, and F. Vinicor, *Circulation* 107, 499 (2003).
[30] A.M. Witkowska and M.H. Borawska, *Eur. Cytokine Netw.* 15, 91 (2004).
[31] P.L. da Luz, P.J. Bertini, and D. Favarato, *Clinics. (Sao Paulo)* 60, 415 (2005).
[32] K.K. Koh, S.H. Han, and M.J. Quon, *J. Am. Coll. Cardiol.* 46, 1978 (2005).
[33] P.M. Ridker, C.H. Hennekens, J.E. Buring, and N. Rifai, *N. Engl. J. Med.* 342, 836 (2000).
[34] I. Fridovich, *J. Biol. Chem.* 272, 18515 (1997).
[35] R.M.T ouyz, *Braz. J. Med. Biol. Res.* 37, 1263 (2004).
[36] D. Han, F. Antunes, R. Canali, D. Rettori, and E. Cadenas, *J. Biol. Chem.* 278, 5557 (2003).
[37] F.Q. Schafer and G.R. Buettner, *Free Radic. Biol. Med.* 30, 1191 (2001).
[38] H.J. Forman and M Torres, *Am. J. Respir. Crit Care Med.* 166, S4-S8 (2002).
[39] U. Landmesser and D.G Harrison, *Coron. Artery Dis.* 12, 455 (2001).
[40] G. Zalba, J.G. San, M.U. Moreno, M.A. Fortuno, A. Fortuno, F.J. Beaumont, and J. Diez, *Hypertension* 38, 1395 (2001).
[41] U. Landmesser and D.G. Harrison, *Circulation* 104, 2638 (2001).
[42] I.H. Schulman, M.S. Zhou, and L. Raij, *Curr. Hypertens. Rep.* 7, 61 (2005).
[43] M. Ushio-Fukai and R.W.Alexander, *Mol. Cell Biochem.* 264, 85 (2004).
[44] S.K. Clinton, R. Underwood, L. Hayes, M.L. Sherman, D.W. Kufe, and P. Libby, *Am. J. Pathol.* 140, 301 (1992).
[45] V.C. Mehra, V.S. Ramgolam, and J.R. Bender, *J. Leukoc. Biol.* 78, 805 (2005).
[46] K. Skalen, M. Gustafsson, E.K. Rydberg, L.M. Hulten, O. Wiklund, T.L. Innerarity, and J. Boren, *Nature* 417, 750 (2002).

[47] S.D. Cushing, J.A. Berliner, A.J. Valente, M.C. Territo, M. Navab, F. Parhami, R. Gerrity, C.J. Schwartz, and A.M. Fogelman, *Proc. Natl. Acad. Sci. U. S. A* 87, 5134 (1990).
[48] T.B. Rajavashisth, A. Andalibi, M.C. Territo, J.A. Berliner, M. Navab, A.M. Fogelman, and A.J. Lusis, *Nature* 344, 254 (1990).
[49] M.T. Quinn, S. Parthasarathy, L.G. Fong, and D. Steinberg, *Proc. Natl. Acad. Sci. U. S. A* 84, 2995 (1987).
[50] T. Collins and M.I. Cybulsky, *J. Clin. Invest* 107, 255 (2001).
[51] W.J. Bettger, *Can. J. Physiol Pharmacol.* 71, 721 (1993).
[52] F. Candan, F. Gultekin, and F. Candan, *Cell Biochem. Funct.* 20, 95 (2002).
[53] M. Demirci, N. Delibas, I. Altuntas, F. Oktem, and Z. Yonden, *J. Health Popul. Nutr.* 21, 72 (2003).
[54] A.M. Roussel, A. Kerkeni, N. Zouari, S. Mahjoub, J.M. Matheau, and R.A. Anderson, *J. Am. Coll. Nutr.* 22, 316 (2003).
[55] A.A. Dimitrova, D.S. Strashimirov, A.L. Russeva, P. Andreeva-Gateva, E.T. Lakova, and K.N. Tzachev, *Folia Med. (Plovdiv.)* 47, 42 (2005).
[56] A. Goel, V. Dani, and D.K. Dhawan, *Chem. Biol. Interact.* 156, 131 (2005).
[57] J.H. Kagi and A. Schaffer, *Biochemistry* 27, 8509 (1988).
[58] B. Bao, A.S. Prasad, F.W. Beck, J.T. Fitzgerald, D. Snell, G.W. Bao, T. Singh, and L.J. Cardozo, *Am. J. Clin. Nutr.* 91, 1634 (2010).
[59] A.S. Prasad, B. Bao, F.W. Beck, and F.H. Sarkar, *Nutrition* (2010).
[60] S. Blankenberg, S. Barbaux, and L. Tiret, *Atherosclerosis* 170, 191 (2003).
[61] L.M. Biasucci, G. Liuzzo, G. Fantuzzi, G. Caligiuri, A.G. Rebuzzi, F. Ginnetti, C.A. Dinarello, and A. Maseri, *Circulation* 99, 2079 (1999).
[62] P.M. Ridker, N. Rifai, M. Pfeffer, F. Sacks, S. Lepage, and E. Braunwald, *Circulation* 101, 2149 (2000).
[63] A. Mazzone, C. Cusa, I. Mazzucchelli, M. Vezzoli, E. Ottini, S. Ghio, G. Tossini, R. Pacifici, and P. Zuccaro, *Clin. Chem. Lab Med.* 39, 822 (2001).
[64] N. Takeuchi, T. Kawamura, A. Kanai, N. Nakamura, T. Uno, T. Hara, T. Sano, N. Sakamoto, Y. Hamada, J. Nakamura, and N Hotta, *Diabet. Med.* 19, 57 (2002).
[65] J.T. Parissis, K.F. Venetsanou, D.G. Mentzikof, M.V. Kalantzi, M.V. Georgopoulou, N. Chrisopoulos, and S.M. Karas, *Eur. J. Intern. Med.* 12, 350 (2001).

[66] R.A. Preston, M. Ledford, B.J. Materson, N.M. Baltodano, A. Memon, and A. Alonso, *J. Hypertens.* 20, 871 (2002).
[67] L. Calabresi, M. Gomaraschi, B. Villa, L. Omoboni, C. Dmitrieff, and G. Franceschini, *Arterioscler. Thromb. Vasc. Biol.* 22, 656 (2002).
[68] A. Hackman, Y. Abe, W. Insull, Jr., H. Pownall, L. Smith, K. Dunn, A.M. Gotto, Jr., and C.M. Ballantyne, *Circulation* 93, 1334 (1996).
[69] G. Lupattelli, R. Lombardini, G. Schillaci, G. Ciuffetti, S. Marchesi, D. Siepi, and E. Mannarino, *Am. Heart J.* 140, 521 (2000).
[70] T. Ohta, K. Saku, K. Takata, and N. Adachi, *Eur. J. Pediatr.* 158, 592 (1999).
[71] C. Bickel, H.J. Rupprecht, S. Blankenberg, C. Espinola-Klein, G Rippin, G. Hafner, J. Lotz, W. Prellwitz, and J. Meyer, *Int. J. Cardiol.* 82, 25 (2002).
[72] A.D. Blann, D. Gurney, E. Hughes, P. Buggins, S.H. Silverman, and G.Y. Lip, *Am. J. Cardiol.* 88, A7 (2001).
[73] E. Oger, M. henc-Gelas, Plu-Bureau, L. Mennen, M. Cambillau, L. Guize, Y. Pujol, and P. Scarabin, *Thromb. Res.* 101, 35 (2001).
[74] P.Y. Scarabin, M. henc-Gelas, E. Oger, and Plu-Bureau, *Thromb. Haemost.* 81, 673 (1999).
[75] D. Zanger, B.K. Yang, J. Ardans, M.A. Waclawiw, G. Csako, L.M. Wahl, and R.O. Cannon, III, *J. Am. Coll. Cardiol.* 36, 1797 (2000).
[76] P.J. Barnes, *Int. J. Biochem. Cell Biol.* 29, 867 (1997).
[77] N.D. Perkins, *Int. J. Biochem. Cell Biol.* 29, 1433 (1997).
[78] A.S. Prasad, F.W. Beck, S.M. Grabowski, J. Kaplan, and R.H. Mathog, *Proc. Assoc. Am. Physicians* 109, 68 (1997).
[79] U. Siebenlist, G. Franzoso, and K. Brown, *Annu. Rev. Cell Biol.* 10, 405 (1994).
[80] D.M. Perkins and J.R. Duncan, *Prostaglandins Leukot. Essent. Fatty Acids* 43, 43 (1991).
[81] A.S. Baldwin, Jr., *Annu. Rev. Immunol.* 14, 649 (1996).
[82] M. Otsuka, M. Fujita, T. Aoki, S. Ishii, Y. Sugiura, T. Yamamoto, and J. Inoue, *J. Med. Chem.* 38, 3264 (1995).
[83] A.S. Prasad, B. Bao, F.W. Beck, and F.H. Sarkar, *J. Lab Clin. Med.* 138, 250 (2001).
[84] U. Zabel, R. Schreck, and P.A. Baeuerle, *J. Biol. Chem.* 266, 252 (1991).
[85] P. Connell, V.M. Young, M. Toborek, D.A. Cohen, S. Barve, C.J. McClain, and B Hennig, *J. Am. Coll. Nutr.* 16, 411 (1997).
[86] E. Ho, N. Quan, Y.H. Tsai, W. Lai, and T.M. Bray, *Exp. Biol. Med. (Maywood.)* 226, 103 (2001).

[87] E. Ho, C. Courtemanche, and B.N.Ames, *J. Nutr.* 133, 2543 (2003).
[88] R.G. Uzzo, P. Leavis, W. Hatch, V.L. Gabai, N. Dulin, N. Zvartau, and V.M. Kolenko, *Clin. Cancer Res.* 8, 3579 (2002).
[89] K. Heyninck and R. Beyaert, *FEBS Lett.* 442, 147 (1999).
[90] M. Jaattela, H. Mouritzen, F. Elling, and L. Bastholm, *J. Immunol.* 156, 1166 (1996).
[91] H.Y. Song, M. Rothe, and D.V. Goeddel, *Proc. Natl. Acad. Sci. U. S. A* 93, 6721 (1996).
[92] R. Beyaert, K. Heyninck, and H.S. Van, *Biochem. Pharmacol.* 60, 1143 (2000).
[93] S.G. Hymowitz and I.E. Wertz, *Nat. Rev. Cancer* 10, 332 (2010).
[94] B.A. Malynn and A. Ma, *J. Exp. Med.* 206, 977 (2009).
[95] F. Blaschke, Y. Takata, E. Caglayan, R.E. Law, and W.A Hsueh, *Arterioscler. Thromb. Vasc. Biol.* 26, 28 (2006).
[96] C. Jiang, A.T. Ting, and B. Seed, *Nature* 391, 82 (1998).
[97] M. Ricote, A.C. Li, T.M. Willson, C.J. Kelly, and C.K. Glass, *Nature* 391, 79 (1998).
[98] T.L. Graham, C. Mookherjee, K.E. Suckling, C.N. Palmer, and L. Patel, *Atherosclerosis* 181, 29 (2005).
[99] S.H. Han, M.J. Quon, and K.K. Koh, *Hypertension* 46, 1086 (2005).
[100] A.C. Li, K.K. Brown, M.J. Silvestre, T.M. Willson, W. Palinski, and C.K. Glass, *J. Clin. Invest* 106, 523 (2000).
[101] A.C. Li, C.J. Binder, A. Gutierrez, K.K. Brown, C.R. Plotkin, J.W. Pattison, A.F. Valledor, R.A. Davis, T.M. Willson, J.L. Witztum, W. Palinski, and C.K. Glass, *J. Clin. Invest* 114, 1564 (2004).
[102] Z. Israelian-Konaraki and P.D. Reaven, *Cardiol. Rev.* 13, 240 (2005).
[103] P. Delerive, B.K. De, S. Besnard, B.W. Vanden, J.M. Peters, F.J. Gonzalez, J.C. Fruchart, A. Tedgui, G. Haegeman, and B. Staels, *J. Biol. Chem.* 274, 32048 (1999).
[104] P. Meerarani, G. Reiterer, M. Toborek, and B. Hennig, *J. Nutr.* 133, 3058 (2003).
[105] G. Reiterer, M. Toborek, and B. Hennig, *J. Nutr.* 134, 1711 (2004).
[106] H. Shen, E. Oesterling, A. Stromberg, M. Toborek, R. MacDonald, and B. Hennig, *J. Am. Coll. Nutr.* 27, 577 (2008).
[107] A. Ashkenazi and V.M. Dixit, *Science* 281, 1305 (1998).
[108] K.A. Lindstedt, M.J. Leskinen, and P.T. Kovanen, *Arterioscler. Thromb. Vasc. Biol.* 24, 1350 (2004).
[109] M.L. Bochaton-Piallat, F. Gabbiani, M. Redard, A. Desmouliere, and G. Gabbiani, *Am. J. Pathol.* 146, 1059 (1995).

[110] W. Cai, B. Devaux, W. Schaper, and J. Schaper, *Atherosclerosis* 131, 177 (1997).
[111] M.M. Kockx, Arterioscler. Thromb. Vasc. Biol. 18, 1519 (1998).
[112] G. Lizard, M. Moisant, C. Cordelet, S. Monier, P. Gambert, and L. Lagrost, *J. Pathol.* 183, 330 (1997).
[113] P. Meerarani, P. Ramadass, M. Toborek, H.C. Bauer, H. Bauer, and B. Hennig, *Am. J. Clin. Nutr.* 71, 81 (2000).
[114] S. Treves, P.L. Trentini, M. Ascanelli, G. Bucci, and V.F. Di, *Exp. Cell Res.* 211, 339 (1994).
[115] F.W. Sunderman, Jr., *Ann. Clin. Lab Sci.* 25, 134 (1995).
[116] Y. Masuda, N. Kawazoe, S. Nakajo, T. Yoshida, Y. Kuroiwa, and K. Nakaya, *Leuk. Res.* 19, 549 (1995).
[117] T. Aiuchi, S. Mihara, M. Nakaya, Y. Masuda, S. Nakajo, and K. Nakaya, *J. Biochem.* 124, 300 (1998).
[118] D.K. Perry, M.J. Smyth, H.R. Stennicke, G.S. Salvesen, P. Duriez, G.G. Poirier, and Y.A. Hannun, *J. Biol. Chem.* 272, 18530 (1997).
[119] Y.J. Geng, Q. Wu, M. Muszynski, G.K. Hansson, and P. Libby, *Arterioscler. Thromb. Vasc. Biol.* 16, 19 (1996).
[120] E.G. Lee, D.L. Boone, S. Chai, S.L. Libby, M. Chien, J.P. Lodolce, and A. Ma, *Science* 289, 2350 (2000).
[121] Y.J. Yang, B.Y. Choi, B.Y. Chun, S.S. Kweon, Y.H. Lee, P.S. Park, and M.K. Kim, *Br. J. Nutr.* 104, 1202 (2010).
[122] P. Connell, V.M. Young, M. Toborek, D.A. Cohen, S. Barve, C.J. McClain, and B. Hennig, *J. Am. Coll. Nutr.* 16, 411 (1997).
[123] B. Hennig, Y. Wang, S. Ramasamy, and C.J. McClain, *J. Nutr.* 122, 1242 (1992).
[124] B. Hennig, Y. Wang, S. Ramasamy, and C.J. McClain, *J. Nutr.* 123, 1003 (1993).
[125] B. Hennig, M. Toborek, C.J. McClain, and J.N. Diana, *J. Am. Coll. Nutr.* 15, 345 (1996).
[126] B. Hennig, P. Meerarani, P. Ramadass, M. Toborek, A. Malecki, R. Slim, and C.J. McClain, *Nutrition* 15, 744 (1999).
[127] C.L. Hou, W. Zhang, Y. Wei, J.H. Mi, L. Li, Z.H. Zhou, W. Zeng, and D.J. Ying, *Genet. Mol. Res.* 10, 1050 (2011).
[128] G.A. Eby and W.W. Halcomb, *Med. Hypotheses* 66, 169 (2006).
[129] D. Steinberg, *Nat. Med.* 8, 1211 (2002).
[130] G. Reiterer, R. MacDonald, J.D. Browning, J. Morrow, S.V. Matveev, A. Daugherty, E. Smart, M. Toborek, and B. Hennig, *J. Nutr.* 135, 2114 (2005).

[131] M. Bhaskar, E. Madhuri, S.A. bdul Latheef, and G. Subramanyam, *Indian J. Exp. Biol.* 39, 1170 (2001).

[132] A. Jenner, M. Ren, R. Rajendran, P. Ning, B.T. Huat, F. Watt, and B. Halliwell, *Free Radic. Biol. Med.* 42, 559 (2007).

[133] M. Bonham, J.M. O'Connor, L.B. McAnena, P.M. Walsh, C.S. Downes, B.M. Hannigan, and J.J. Strain, *Biol. Trace Elem. Res.* 93, 75 (2003).

[134] S. Hercberg, S. Bertrais, S. Czernichow, N. Noisette, P. Galan, A. Jaouen, J. Tichet, S. Briancon, A. Favier, L. Mennen, and A.M. Roussel, *Lipids* 40, 335 (2005).

[135] E. Ari, Y. Kaya, H. Demir, E. Asicioglu, and S. Keskin, *Biol. Trace Elem. Res.* (2011).

[136] M. Ghayour-Mobarhan, A. Taylor, S.M. Kazemi-Bajestani, S. Lanham-New, D.J. Lamb, N. Vaidya, C. Livingstone, T. Wang, and G.A. Ferns, *Clin. Lab* 54, 321 (2008).

[137] G.D. Giannoglou, D.M. Konstantinou, L. Kovatsi, Y.S. Chatzizisis, and D.P. Mikhailidis, *Angiology* 61, 449 (2010).

[138] Y. Islamoglu, O. Evliyaoglu, E. Tekbas, H. Cil, M.A. Elbey, Z. Atilgan, H. Kaya, Z. Bilik, A. Akyuz, and S. Alan, *Biol. Trace Elem. Res.* (2011).

[139] S.M. Kazemi-Bajestani, M. Ghayour-Mobarhan, M. Ebrahimi, M. Moohebati, H.A. Esmaeili, M.R. Parizadeh, R. Aghacizadeh, and G.A. Ferns, *J. Trace Elem. Med. Biol.* 21, 22 (2007).

[140] D.H. Lee, A.R. Folsom, and D.R. Jacobs, Jr., *Am. J. Clin. Nutr.* 81, 787 (2005).

In: Zinc: Characteristics, Uses and Benefits
Editor: Masayoshi Yamaguchi
ISBN: 978-1-61942-523-1

Chapter 5

NEW INSIGHTS INTO THE ROLE OF ZINC IN CARDIOPROTECTION DURING ISCHEMIA/REPERFUSION INJURY

***Vijaya Lakshmi Bodiga*[1*] *and Sreedhar Bodiga*[2]**

[1] Department of Biotechnology, Krishna University, Machilipatnam, Andhra Pradesh, India.

[2] Division of Cardiology, Department of Medicine, Cardiovascular Research Center, University of Alberta, Edmonton, AB Canada.

ABSTRACT

Zinc has recently emerged as an important trace element, capable of conferring cardioprotection and a variety of other benefits in the cardiovascular system and elsewhere. The events that occur subsequent to coronary artery occlusion are termed myocardial ischemia/reperfusion denoting that these are distinct phases of cellular injury with ATP depletion, lactate accumulation and acidosis observed during ischemia and production of reactive oxygen species during reperfusion. Cell death during myocardial ischemia/reperfusion occurs by necrosis and apoptosis, both regulated by stress signals and results in "cardiac wasting." Although the mechanisms by which zinc accomplishes cardiac benefits remain largely unclear, proposals include both direct and indirect effects on an array of intracellular targets such as adenosine receptor, ErbB2, NADPH oxidase, PKC, Akt, GSK-3β and mPTP culminating in

---

* E-mail: Vijaya Bodiga <bodigavijayasri@gmail.com>

anti-apoptotic effects and improved functional recovery of contractile function. Interestingly, cardioprotection by zinc is manifested only in the presence of a lipophilic ionophore, such as pyrithione, in isolated neonatal and adult rat cardiac myocytes subjected to simulated ischemia/reperfusion, Langendorff isolated rat heart perfusion and ischemia/reperfusion injury in myocardial infarction. Thus, although great potential exists for zinc-based therapeutics, a number of questions remain. This review will discuss the available literature pertaining to zinc homeostasis in the myocardium and provide novel insights into the underlying molecular mechanisms towards better understanding of the therapeutic effect of zinc during ischemia/reperfusion. Attempts are made to identify the gaps in basic and translational science in the development of zinc ionophores as potential therapeutic compounds in the setting of myocardial ischemia reperfusion injury.

INTRODUCTION

Zinc Distribution in Myocardium and Functions

Zinc is one of the most abundant transition metals in the human body, involved in a myriad of biological functions [1]. It is an essential structural and regulatory component of many proteins and enzymes, including transcription factors and various signaling proteins with zinc-binding motifs. In fact, zinc is also known to stabilize protein-protein interactions and complex formation [2]. Hence, it is not surprising that our body contains a substantial amount of zinc in every cell and is the second most abundant trace element in the body. Extracellular free zinc reaches micromolar concentrations, while free intracellular Zn^{2+} concentrations range from picomolar to nanomolar. Such a steep electrochemical gradient may even exceed the transmembrane potential of Ca^{2+} gradient and requires an active mechanism that extrudes extracellular zinc. Such a mechanism is essential for regulating intracellular zinc and protecting against excessive extracellular zinc. Further, the total zinc concentration in eukaryotic cells is ~ 200 μM, whereas cytoplasmic free zinc concentrations are only in pM to nM concentrations. In quiescent/resting cardiomyocytes, levels below 100 and 520 pM were found [3, 4]. The cardiac tissue contains moderate levels of zinc compared to rest of the body. A gram of mouse heart tissue contains 26-35 μg of total zinc, but is not homogeneously distributed [5]. In particular, the right ventricle contains higher levels of zinc than does the rest of the heart tissue, possibly reflecting the higher metabolic turnover in the left ventricle as it has to pump against a 4-

6~ fold high pressure than the right ventricle. Intracellular zinc is found in tight association with metalloproteins, a largely fixed pool with a slow turnover, or in a more dynamic, free or loosely bound form (labile) that is more readily exchangeable. The amount of intracellular labile Zn differs according to cell type and can participate in specialized function such as secretion, cytoprotection and signal transduction [6]. Most of the zinc in the heart, like in the rest of the body, is tightly bound to proteins, and thus inaccessible to simple histochemical detection methods. However, a substantial fraction of the zinc is present in chelatable (free or loosely bound) form and hence can be stained with N-(6-methoxy-8-quinolyl)-p-toluene suphonamide (TSQ) and other fluorophores. Although proton/zinc exchange activity has been demonstrated in plasma membrane vesicles derived from brain cells [7], it seems unlikely that the small magnitude of the proton electrochemical gradient is sufficient to catalyze the formation of steep Zn^{2+} gradients across the plasma membrane of other mammalian cells. Furthermore, other studies have suggested that zinc allosterically inhibits cation transport mediated by Na^{+}- Ca^{2+} exchanger [8]. The ZnT-1 is yet another protein suggested as a mediator of zinc efflux, and its expression was shown to confer resistance against zinc toxicity and to lower intracellular zinc [9,10]. Very recent studies provide evidence for a novel Na^{+}/Zn^{2+} exchanger which utilizes the electrochemical Na^{+} gradient, to catalyze the active extrusion of intracellular zinc. Activity of Na^{+}/Zn^{2+} exchange is demonstrated by the Na^{+}-dependent extrusion of zinc against a 500-fold zinc gradient and is not influenced by ATP depletion [11]. Various observations of zinc ion fluctuations in the heart support the notion that intracellular zinc signals are used more widely. When subcellular distribution of zinc was investigated in cardiomyocytes, transverse striations of millimolar concentrations (10^{-4} and 10^{-3}M) of zinc were found with a periodicity of sarcomeres and corresponding to the I-band. It was suggested that these intracellular zinc stores can be released into the cytosol within seconds [12]. Systemically, plasma zinc levels are ~ 1 μg/ml, of which 67% is mobile, transportable zinc available for uptake by cells [13]. Zinc content in the serum and/or heart is abnormally low in humans with heart failure or other conditions such as diabetes, hypertension, coronary artery disease or myocardial infarction [14-17]. In myocardial infarction, serum zinc levels fall within one or two days of the onset, and then rise to reach normal levels in 10-14 days. These variations have been related to alterations in enzymatic markers of this pathology. Amazingly, cardiac zinc content positively correlates with ejection fraction in humans with coronary artery disease [18]. Myocardial zinc levels in albino rats administered isoproterenol to induce an acute hyperadrenergic state and cardiac necrosis showed a significant decrease and correlated with the extent of necrosis [19,20]. Elevated cardiac zinc status also preserves cardiac function in the face of oxidative stress associated with ischemia/reperfusion or diabetes. Hyperzincuria coupled with hypozincemia have been consistently associated with congestive heart failure [21-24]. In the heart, zinc is also known to affect differentiation and regeneration of cardiac muscle [25], cardiac conductance, acute stress responses and recovery of heart transplants [26-27]. Thus, it is not only an essential micronutrient and growth factor, but is involved in structural and functional integrity of the myocardium. Interestingly, one of the hallmarks of zinc deficiency is cardiac dysfunction [28]. These paradoxical aspects of cellular zinc dictate that it be distributed in highly regulated gradients with respect to the plasma membrane and intracellular compartments. It also demands that an adequate supply of this ion be available to the numerous zinc-binding proteins, while preventing its accumulation and the potentially devastating effects it can readily initiate. This emphasizes the pertinent role of zinc in the myocardium in experimental models as well as in clinical studies, and suggests its involvement in many critical cellular functions.

Myocardial Ischemia and Reperfusion (I/R)

Myocardial ischemia refers to reduced coronary blood flow to myocardium, either absolute flow rate (low-flow or no-flow ischemia) or relative to increased tissue demand (demand ischemia). A pivotal feature of ischemia is lack of metabolic nutrients and diminished oxygen supply to the mitochondria to support oxidative phosphorylation [29,30]. In experimental models and clinical situations, ischemia may be followed by reperfusion, that is, the re-entry of oxygen and metabolic substrates with removal/elimination of ischemic metabolites. The process of reperfusion is associated with oxidative burst and altered biochemical, structural, and functional changes in myocardium and may determine cell survival and cell death. Cell death following I/R has been reported to have features of apoptosis, autophagy, and necrosis. The precise proportion of each form of cell death may depend on the model. Apoptotic cell death results in apoptotic bodies that contain cellular components and these apoptotic bodies are phagocytized by other cells and is generally characterized by chromatin condensation, fragmentation, cell shrinkage, and plasma membrane budding with release of apoptotic bodies that are phagocytized. Apoptotic cell death typically occurs via activation of caspases that cleave DNA and other cell components. Deletion of pro-apoptotic proteins or increased expression of anti-apoptotic proteins have been shown to reduce I/R-mediated cell death. Autophagy is another process that has been suggested to play a role in I/R injury. It is a physiological mechanism used to remove damaged organelles, such as mitochondria or endoplasmic reticulum. This process is generally initiated by starvation to provide nutrients. The role of autophagy in cell death during I/R is still somewhat unclear and debated as to whether is beneficial or detrimental. Necrotic cell death is characterized by cell swelling leading to irreversible rupture of the plasma membrane with release of cytosolic components, resulting in an inflammatory response. The release of troponin and creatine kinase that occur during I/R are likely due to necrosis. Most importantly, the dynamic regulation of metabolic and electrophysiological processes becomes dysfunctional, often resulting in alterations in ionic balance and the abrupt loss of cardiac function in the form of fatal cardiac arrhythmias and/or contractile failure. TSQ staining in cryosections of normally perfused heart revealed Zn^{2+} presence in sarcomeric units, filaments, perinuclear area, and intercalated discs, but not in nuclei. Interestingly, TSQ staining for Zn^{2+} varied dramatically in the hearts subjected to global ischemia followed by reperfusion with a significantly decline in the total Zn^{2+}, and confined to the zones of cell contacts, as well as patchy areas containing granulated vesicle-like structures, termed 'zincosomes' [31]. Therefore, despite the recently emerging experimental evidence, the role of intracellular zinc and its mobilization within the cell remains poorly understood. Repeated studies consistently show that the zinc levels of cardiomyocytes and myocardial tissue after I/R are lower than the normal myocardial tissue. This consistency persists in different reports by different investigators employing different populations and tissue samples and involving the use of various fluorophores for assessment of zinc levels. Thus it appears that impaired zinc accumulation in the myocardium critically relates to ischemia reperfusion injury and if such is the case, one should expect that the zinc-accumulating process is absent or defective in the myocardium during I/R and necessitates the use of ionophores to transport zinc across the lipid membrane. It is common to consider that ischemia/reperfusion (I/R) injury is a composite entity with distinct components of injury associated specifically with ischemia and with reperfusion since it is a truism that reperfusion can never occur independently of ischemia. Therefore, elucidation of the mechanisms underlying I/R has proved to be no simple task. It was initially believed that the cellular damage observed after restoration of blood flow to ischemic tissue was caused solely by reactive oxygen species. Although studies have indicated that ROS do play a role in I/R, it has since been demonstrated that there are several other mediators involved in the pathogenesis of I/R injury, which include nitric oxide, inflammatory cells, cytokines, eicosanoids, the complement cascade, endothelial cells and adhesion molecules. Impaired ionic homeostasis also significantly contributes to I/R injury. The interactions among these mediators are complex and are yet to be fully revealed. An additional complication in studying I/R injury is the observation that the mechanism and manifestation of injury varies somewhat with the model being used for such studies. Although a complete review of the mechanisms underlying I/R injury is beyond the scope of this chapter, we will address some of the principal mediators of injury and focus on targets that are known to be affected by zinc and offer a clinical promise as a new therapeutic intervention to limit I/R injury. Thus, myocardial ischemia and reperfusion (I/R) is a complex set of events that are associated with significant alterations in cellular metabolism, oxidant stress, and triggering of pro-inflammatory molecules [32]. As oxygenated blood is supplied to ischemic tissue, myocardium is paradoxically predisposed to further injury, ultimately leading to larger infarcts and impaired ventricular function [33]. Although many agents have been used experimentally to curtail I/R injury, few have proven clinical efficacy [34,35]. Also, since pretreatment is impossible in the clinical setting of acute myocardial infarction, the use of the pre-ischemic interventions is clinically impractical. Thus, to successfully save myocardium in patients with acute myocardial infarction, protective interventions must be effective when applied after ischemia or at the onset of reperfusion. Thus, identifying effective and clinically accessible methods of limiting myocardial injury associated with I/R remains an important and challenging issue.

Zinc Homeostasis and Ionophores

Any compound capable of binding zinc with moderate affinity and having sufficient lipophilic properties to penetrate cell membranes is capable of effecting intracellular zinc concentrations can be called a zinc ionophore. Various compounds such as pyrithione, heterocyclic amines such as 5,7-diiodo-8-hydroxyquinoline, and 8-hydoxyquinoline, the dithiocarbamates, including pyrrolidine dithiocarbamate and diethyldithiocarbamate, disulfiram and dimethyldithiocarbamate and vitamins including vitamin E and vitamin A are all endowed with lipophilic as well as zinc binding properties and hence are useful in manipulating intracellular zinc levels. Zinc pyrithione (zinc pyridinethione, $C_{10}H_8N_2O_2S_2Zn$, MW 317.75) is highly lipophilic and therefore penetrates membranes easily, facilitating the transport of zinc across cell membranes. Use of zinc pyrithione or other zinc ionophores such as dithiocarbamates is necessitated by the fact that all eukaryotic cells strictly regulate the membrane transport of Zn^{2+}, making it difficult to modulate the intracellular concentration and distribution of zinc. Until now, no information is available regarding the specific mechanisms of zinc import and export in cardiomyocytes. Although oxidants like H_2O_2 enhance $[Zn^{2+}]_i$ by roughly 30-fold [36], this increase does not afford protection in ischemic injury. However, this free Zn^{2+} can contribute to oxidant-induced alterations of excitation-contraction coupling [36]. In addition, Zn^{2+} release, a phenomenon termed 'Zn^{2+} wave', originating from the endoplasmic reticulum and/or the surrounding area in mast cells after cross-linking of the high-affinity immunoglobulin E receptor, was shown to be dependent on Ca^{2+} influx without any contribution of extracellular Zn^{2+} [37]. Alternatively, free Zn^{2+} can also be sequestered into cellular organelles, a mechanism that complements physico-chemical buffering and has been referred to as 'muffling' [38]. In order to maintain controlled levels and appropriate distribution of highly charged metal ion, tightly regulated homeostatic mechanisms need to be in place. This is achieved through the action of a set of specialized proteins that control Zn uptake, efflux and compartmentalization. These proteins and their genes are being rapidly identified and characterized [39,40]. Solute Carrier Family 39A (SLC39A) includes mammalian ZRT/IRT-related proteins (ZIPs) [41], and Solute Carrier Family 30A (SLC30A), comprises mammalian ZnTs [42]. Both ZIPs and ZnTs are multipass transmembrane proteins and have a high content of lipophilic amino acids. Fourteen ZIP-encoding genes have been identified in the human genome and are grouped into four sub-families according to the molecular features of the encoded proteins. All of the Zip proteins mediate Zn uptake from the extracellular environment or intracellular vesicles into the cytoplasm. Proteins belonging to the ZnT family were shown to be responsible for Zn efflux from the cytoplasm towards either intracellular vesicles and/or the extracellular space [43], as wells for Zn delivery to specific metalloproteins [44]. At least 3 distinct isoforms of metallothionein aid in buffering of intracellular zinc [45]. Metallothionein (MT) is a cysteine-rich 6-7 kDa protein that binds metal ions, in particular zinc [46]. Although all zinc ions are bound in similar tetrathiolate coordination environment, the affinities of the sites for zinc vary, with a logK between 7.7 and 11.8 [47], providing picomolar to nanomolar concentrations of free zinc ions to the cytosol. Zinc thiolate clusters in metallothionein bind

Zn^{2+} rather tightly and yet can make it available, providing a mechanism whereby Zn^{2+} availability and redox metabolism are linked [46]. Zn^{2+} can be released from MT in response to disulfide containing oxidants, such as oxidized glutathione (GSSG) or 2,2'-dithiopyridine, leading to the suggestions that metallothionein-zinc pool might be mobilized in response to changes in the redox potential of the cell [48]. However, it is not known if the mitochondrial and cytosolic Zn^{2+} pools can be mobilized largely independent of each other and the fate of zinc released due to intracellular oxidative stress has not been studied in detail. Interestingly, Kang et al. used a Langendorff-perfusion mouse heart model and demonstrated that metallothionein (MT) functions in myocardial protection against ischemia-reperfusion injury [49]. Because MT can both bind to and be induced by heavy metal ions such as Cu, Cd, Co, Hg and Zn, it is generally argued that MT functioning as an antioxidant is the effector in cytoprotection in the heart, due to suppression of oxidative damage, although the mobilization of zinc from metallothionein cannot be ruled out. Using cardiac –specific, MT-overexpressing, transgenic mouse model, it was demonstrated that elevated MT (~10-fold higher) significantly improved the suppressed contractile force post-ischemia and reduced the efflux of creatine kinase by more than 50%. In addition, the zone of infarction induced by I/R was suppressed by 40% in the MT-overexpressing transgenic heart, strongly suggesting the capacity of MT to protect against ischemia-reperfusion induced heart injury [49,50]. Based on the available literature, it can be suggested that zinc release can induce translocation and activation of Zn^{2+}-sensitive transcription factors (MTF-1) to induce the expression of certain genes like MT-1 and MT-2 [51]. This does mean that ability of zinc to induce metallothionein can be a contributing factor in protecting against myocardial ischemia reperfusion. Maintenance of body and cellular Zn homeostasis therefore requires a great number of genes whose polymorphisms and mutations are likely to provide complex genetic variations and sensitivity to dietary Zn in health and disease. Among various zinc transporters, expression of ZnT-1, ZnT-5 and ZnT-7 has been observed in heart [52]. Mitochondria constitute another likely site of intracellular Zn^{2+} sequestration. Elevated cytosolic Zn^{2+} forces the mitochondria to take up Zn^{2+}, from which it can be subsequently released in a Ca^{2+} dependent fashion [53,54]. Zn^{2+} accumulation in the mitochondria may not be benign, as Zn^{2+} has been found to interfere with the function of isolated mitochondria [55-62].

Cardioprotective Effects of Zinc during I/R

Zinc salts and ionophores are emerging as therapeutics in cardioprotection and in improving cardiac function through their effects on Ca^{2+} regulation, myofilament force-production and/or proximal second messengers. Studies by Powell et al. (1994) documented the protective effect of zinc ions in myocardial recovery from I/R. The study used zinc-bis-hisidinate to improve membrane permeability of the zinc. Addition at pre-ischemia conferred myocardial protection, whereas treatment starting at reperfusion worsened post-ischemic damage, thus limiting clinical applications of zinc. The protective effect was attributed to changes in redox metabolism, namely to decrease in hydroxyl radical formation and copper reactivity [63]. These data promoted renewed interest in cardioprotective roles of zinc. Karagulova et al. (2007) further carried out studies in Langendorff perfused hearts to evaluate the effects of zinc pyrithione [31]. TSQ staining in cryosections of normally perfused heart revealed Zn^{2+} presence in sarcomeric units, filaments, perinuclear area, and intercalated discs, but not in nuclei. Interestingly, TSQ staining for Zn^{2+} varied dramatically in the I/R hearts with a significantly decline in the total Zn^{2+}, and confined to the zones of cell contacts, as well as patchy areas containing granulated vesicle-like structures, termed 'zincosomes'. The distribution of labile zinc in I/R tissues supplemented with zinc pyrithione at the onset of reperfusion was very similar to the normally perfused rat hearts, but showed the highest level of basal zinc. These data for the first time suggested that heart tissue becomes zinc-depleted after I/R and that the capacity to liberate or mobilize labile zinc by a classical trigger phorbol myristyl acetate (PMA) is significantly diminished, indicating zinc dyshomeostasis after I/R. More strikingly, zinc pyrithione was able to restore the basal zinc levels during I/R and prevent apoptosis. In hearts subjected to both reversible and irreversible global ischemia followed by reperfusion, zinc ionophore, pyrithione applied at the time of reperfusion completely recovered left ventricular developed pressure (LVDP) and reduced the incidence of arrhythmias more than ~2-fold. More interesting observation was that the effect of zinc pyrithione was enhanced when the ischemia period was extended up to 30 min, yielding the improvement of LVDP recovery from 15.58% to 74.95% and diminished the incidence of ventricular fibrillation by 8-fold. The dose response of zinc pyrithione treatment was biphasic with the increase of protection at the concentrations up to 10 μM followed by a sharp decrease most likely due to Zn^{2+} toxicity. Other interesting observations from this study were that the recovery of LVDP with zinc pyrithione was attenuated by co-administration of TPEN, a Zn^{2+} chelator and that zinc pyrithione supplementation in the diet could recapitulate the LVDP recovery to the same extent as zinc pyrithione perfusate. The capacity of zinc pyrithione to prevent PKCε reduction and PKCδ cleavage after I/R was remarkable and contributed to the overall anti-apoptotic effect of zinc pyrithione during I/R, although the upstream and downstream signaling events pertaining to the involvement of PKC with zinc remain unclear. Activation of PKCε has been shown in numerous studies to have an important role in cardioprotection. PKCε translocation to the membrane fraction has been used as a measure of activation and this process is blocked by wortmannin, suggesting it is downstream of PI3K [64]. Recent studies have further demonstrated that total zinc concentration as assessed by ICPOES decreases upon reperfusion in the heart. However, exogenous zinc (10 μM) in the presence of pyrithione (4 μM) for 20 min protects cardiac H9c2 cells from reperfusion injury by targeting the mitochondrial permeability transition pore (mPTP) opening via Akt and GSK-3β-Ser9 phosphorylation [65,66]. GSK-3β is phosphorylated and inhibited in a wortmannin-sensitive manner. In unstimulated cells, GSK-3β is active and phosphorylates downstream substrates. But phosphorylation by Akt, p70S6K, PKC, PKA and p90RSK have all been reported to inactivate GSK-3β, resulting in anti-apoptotic effects. There are considerable data suggesting that inhibitors of MPT, a large conductance channel, reduce I/R injury. The MPT has been suggested to be a multiprotein complex composed of the ANT and VDAC and which is modulated by cyclophilin. The MPT opens under conditions of high matrix Ca^{2+}, ROS, high NADH, depletion of adenine nucleotides, and loss of $\Delta\psi$. Contrastingly, low pH, as seen during ischemia inhibits MPT, whereas rapid restoration of intracellular pH during early reperfusion allows activation of MPT. It is thus important to understand that a reduction in cytosolic Ca^{2+} would reduce Ca^{2+} uptake into the mitochondria during I/R and thus reduce MPT. Similarly, a reduction in ROS or inhibition of complex I have been reported to reduce opening of MPT. It was also demonstrated that nitric oxide (NO) prevented mitochondrial oxidant damage by mobilizing intracellular zinc via the cGMP/PKG signaling pathway, providing evidence in favor of the critical role of zinc in preventing myocardial reperfusion injury [67]. Ping et al. showed that NO donors or NO caused translocation of PKCε, suggesting that NO is upstream of PKCε [68]. It is known that eNOS can be phosphorylated and activated by Akt, suggesting a mechanism that would place PKCε downstream of PI3K [69]. The ability of zinc pyrithione to activate Akt and inactivate GSK-3β was dependent on

PI3Kinase and Akt, but not on p70S6 kinase, mTOR or PKC pathway. The action of PI3K is usually attributed to generation of phosphatidylinositol 3,4,5-trisphosphate, that facilitates PDK1 phosphorylation of substrates. PDK1 has been reported to phosphorylate Akt, p70S6K, PKC, PKA, although it is not clear if all of these targets are phosphorylated in response to zinc administration. PIP3 generated by PI3K is thought to bring Akt to the membrane and to alter Akt conformation to facilitate phosphorylation by constitutively active PDK1. It will be interesting to test Yellon's proposal that zinc can act directly on the initiation of the phosphatidylinositol 3-kinae (PI3K)-Akt signaling cascade by inhibiting the negative regulators of this pathway, namely the protein tyrosine phosphatases (PTPases). This proposal stems from the previous observations that zinc ions are known to modify the activity of PTPases by reacting with specific cysteine residues localized in the active site of these enzymes. This reaction inactivates the phosphatase and can trigger the process of ubiquitination and proteasomal degradation [70]. The most important PTPase in the downregulation of PI3K/Akt pathway is the phosphatase and tensin homolog on chromosome 10 (PTEN) [71]. Interestingly, it has already been shown in human airway epithelial cells and rat lungs that treatment with zinc results in a significant reduction of the levels of PTEN, parallel with an increased Akt phosphorylation [72]. In addition to PTEN, but to a lesser extent, other tyrosine protein phosphatases (e.g., Src homology 2-containing inositol phosphatase 2) or protein phosphatases (e.g., PH domain leucine-rich repeat protein phosphatase and protein phosphatase 2A) can modulate the activity of the PI3K/Akt pathway, a pathway that needs strict regulation to avoid initiating uncontrolled tissue growth [73].

Adenosine pretreatment has been shown to reduce acidosis during ischemia and slows down the rate of ATP breakdown [74]. Adenosine can produce NO through A_2 receptors [75] and NO can mobilize intracellular zinc [67]. Therefore, it was thought that intracellular zinc would play a role in A_2 receptor activation induced cardioprotection at reperfusion. Interestingly, adenosine A_1/A_2 receptor agonist, 5'-(N-ethylcarboxamido) adenosine (NECA) given at reperfusion reduced infarct size in isolated rat hearts subjected to 30 min of ischemia followed by 2 h of reperfusion, an effect that could be significantly blocked by zinc chelator, N,N,N',N',-tetrakis-(2-pyridylmethyl) ethylenediamine (TPEN). In addition, NECA increased intracellular zinc suggesting an important role for zinc in NECA-mediated adenosine receptor activation and cardioprotection [76]. It is well known that periods of ischemia are often accompanied by a significant outpouring of adenosine from injured tissue, and this endogenously released adenosine

the nucleus and binds to the κB motif of DNA [106]. It also plays a pivotal role in myocardial IR injury [107]. Ischemia reperfusion leads to the activation of NF-kB, increases the plasma level of pro-inflammatory mediators and enhanced infiltration of neutrophils to the infarct area, resulting in contractile dysfunction and ischemic damage [108,109]. Recently, studies have documented a potent role for the NF-kB in the regulation of cardiac myocyte survival through repression of apoptotic cell death triggered by hypoxia or ischemic myocardial injury [110-112]. These studies strongly suggested a protective role for NF-kB during pathological remodeling of the heart following acute cardiac injury. However, the exact role of the NF-kB in the heart remains a source of controversy, where the role of NF-kB has been described as both adaptive and maladaptive in certain cellular contexts. Indeed, following ischemia and reperfusion in the heart, at least 2 waves of NF-kB activity has been described over the first 6 hours of reperfusion, and recent evidence suggests that NF-kB activity may be sustained for weeks following coronary ligation [108,113]. In this context, chronic activation of NF-kB may induce the expression of inflammatory cytokines and produce detrimental consequences, including cardiac cell death. However, there is no direct evidence so far in proof of this concept that zinc might be acting through the inhibition of proteasome activity or NF-kB inactivation in reperfusion injury.

Despite the general consensus on the role of zinc in myocardial ischemia/ reperfusion injury, relevant questions remain unsolved, especially concerning molecular mechanisms and causal relationships. These new data provide a way to reconcile the results of previous studies of cardioprotection by zinc and provide a better model for understanding zinc homeostasis and one which can be applied to therapeutic purposes.

Modulation of Reactive Oxygen Species (ROS) Levels by Zinc in I/R Injury

Reactive oxygen species (ROS) could, in principle, have several different effects in the heart, although the sources and targets of ROS are still debatable. Traditionally, free radicals have been recognized to be capable of inducing oxidation and damage of macromolecules, membranes and DNA and thereby contributing to cellular damage and acceleration of cell death through apoptosis and necrosis. In addition, ROS can profoundly impair cellular energetics through actions on mitochondrial enzymes. More recently, it has been appreciated that oxidative stress may also exert more subtle modulatory effects on redox signaling. The sources of ROS within myocardial cells include the mitochondrial electron transport chain, xanthine oxidase, uncoupled nitric oxide synthase, cytochrome P450, and NADPH oxidase. NADPH oxidase belongs to the family of oxidoreductases that catalyze the transfer of electrons from the reductant (electron donor) to the oxidant (electron acceptor). Unlike other oxidoreductases, NADPH Oxidase is a distinct enzymatic source of cellular ROS generation [114], whereas the other enzymes produce ROS only as by-products along with their specific catalytic pathways. The NADPH oxidase paradigm was largely derived from studies in phagocytic leukocytes, such as neutrophils and macrophages [115]. The backbone of this enzyme is the membrane-bound cytochrome b558, which consists of two subunits $gp91^{phox}$ (Nox2) and $p22^{phox}$. Upon cell activation, two cytosolic regulatory subunits $p47^{phox}$ and $p67^{phox}$, as well as a small G protein Rac, translocate to the membrane and associate with the cytochrome b448. The assembled multi-subunit enzyme complex then generates superoxide by one-electron reduction of oxygen *via* its $gp91^{phox}$ subunit using reduced NADPH as the electron donor [115]. During the last decade, NADPH oxidase activity was observed in non-phagocytic cells and with this observation, additional isoforms of NADPH Oxidase were discovered. Recent studies have shown that NADPH oxidase is a major source of ROS in cardiomyocytes, endothelial cells, smooth muscle cells and fibroblasts [116-118]. In cardiovascular tissues, Nox2 and Nox4 are expressed in endothelial cells, cardiomyocytes and fibroblasts [119-123,116]. The increased expression of Nox2 is related to cell damage after acute myocardial infarction in rats [124] and $p47^{phox}$ is thought to play a key role in the activation of Nox2 [125]. Nox4 is reported to be constitutively active for basal ROS formation and can also be induced by stimuli such as angiotensin II, insulin and TGF-β1 [126]. In rat myocardial cell line, H9c2, hypoxia (24 h) followed by reoxygenation (12 h) significantly increased Nox2, Nox4 and $p47^{phox}$ expression, NADPH oxidase activation, and ROS production, suggesting the involvement of NADPH oxidase towards ROS production [127]. Given the critical role of $p47^{phox}$ and NADPH oxidase complex in myocardial oxidative stress, and decreased intracellular zinc concentrations during myocardial ischemia and reoxygenation, it was thought relevant that depleted zinc levels at least partially contribute to myocardial damage, due to altered redox balance. In this direction, 1-hydroxy-2-pyridinethione, or 2-mercaptopyridine N-oxide, also known as pyrithione, a lipophilic ionophore and its metal complex, zinc pyrithione was shown to rapidly transport Zn^{2+} into cells. Using 10 μM $ZnCl_2$ and 4 μM pyrithione in primary neonatal rat cardiac myocytes, it was demonstrated that zinc pyrithione effectively attenuates hypoxia/reoxygenation-induced increase in cell death and improved cell viability, as shown by MTT assay and LDH release. Hoechst 33258 staining showed that H/R resulted in increased number of cells with a bright blue fluorescent condensed nuclei and chromatin fragmentation, indicating pyknotic nuclei. Treatment with zinc pyrithione at the onset of reperfusion significantly suppressed the pyknotic nuclei formation and also reduced the DHE fluorescence as well as lucigenin-derived chemiluminescence, indicative of superoxide formation. The DHE fluorescence intensity as well as the lucigenin-derived chemiluminescence signal after H/R was abolished by treatment with $gp91^{phox}$ ds tat peptide inhibitor that blocks the association of $gp91^{phox}$ with $p47^{phox}$, consistent with the role of NADPH oxidase in superoxide generation during H/R. Taqman real-time PCR and Western blotting further confirmed the increase in NOX2 ($gp91^{phox}$) and $p47^{phox}$ mRNA levels after H/R compared to normoxia control. Zinc pyrithione significantly reduced the mRNA and protein levels of NOX2. Most significantly, zinc pyrithione effectively suppressed the H/R-induced increased phosphorylation of $p47^{phox}$ on Ser residue. These data thus suggested that NADPH oxidase is a critical regulator of reperfusion injury and that $p47^{phox}$ activation can be targeted with the use of zinc pyrithione during reperfusion [128].

Another major source of ROS during cardiac ischemia and reperfusion is the mitochondria and thus forms an important target. Ischemia causes progressive damage to mitochondria, including the electron transport chain [129-131]. Complex I activity decreases within 30min [130,132], and affecting complex V [132] and the adenine nucleotide transporter [133,129] relatively early in ischemia. With increasing duration of ischemia, even complex III [134] and cytochrome oxidase activity decrease [135-137]. Thus, while complex I defect occurs early, as ischemia continues, damage progresses to involve complexes III and IV. Complex I activity decreases during ischemia due to a decrease in the NADH dehydrogenase component [138,132], likely due to the loss of flavin mononucleotide coenzyme. Damage to the NADH dehydrogenase component of complex I can increase electron leak and the production of ROS [139]. In addition, ischemia leads to the release of cytochrome c from mitochondria and activates downstream caspase-3 leading to an increase in the number of cardiomyocytes displaying apoptotic markers that become increasingly evident during reperfusion [140-143]. Interestingly, zinc pyrithione administered to neonatal rat cardiac myocytes in culture and subjected to hypoxia/reoxygenation effectively decreased the proportion of cells with apoptotic markers and improved ATP generation [144,128]. It can be argued that mitochondria contribute to their own damage during ischemia, since blockade of electron transport immediately before ischemia dramatically attenuates damage to the mitochondrial electron transport, with preserved oxidative phosphorylation, retention of cytochrome c, and decreased release of ROS. Transient blockade of mitochondria during early reperfusion, by ischemia, hypoxia, or pharmacological inhibition paradoxically decreases myocardial injury, presumably by blunting the deleterious consequences of respiration in mitochondria that suffered ischemic damage. Thus, the restriction of mitochondrial oxidative metabolism by therapeutic intervention via the activation of signaling cascades is a key mechanism of cardiac protection during ischemia and reperfusion. In fact, interaction of zinc with the respiratory chain components was first observed by Skulachev et al. [145], who reported its inhibitory effect on the cytochrome bc_1 complex. Later, Zn^{2+} inhibition of cytochrome *c* oxidase, with 4 redox centers was observed [146,147]. The effects of zinc on mitochondrial NADH:ubiquinone oxidoreductase (complex I) have revealed that Zn^{2+} inhibits complex I more strongly than other metal ions, but does not inhibit NADPH oxidation or intramolecular electron transfer, thus representing a novel complex I inhibitor distinct from the ubiquinone site inhibitors. Zinc probably inhibits either proton transfer to bound quinine or proton translocation [148]. Under physiological conditions, ATP is mainly supplied by mitochondrial oxidative phosphorylation and to a lesser degree by glycolysis. Commensurate with the large metabolic demand of contraction of the heart, mitochondria occupy approximately 30% of the volume of ventricular myocytes [149], and form a network around the myofilaments, thus placing the sites of ATP production immediately adjacent to the site of ATP consumption [150]. The critical dependence of the ion homeostasis on sufficient energy supply becomes evident in pathophysiological conditions as diverse as ischemia/reperfusion, heart failure, and ventricular hypertrophy, in which a mismatch in ATP supply and ATP utilization may lead to electric and mechanical instability. During severe ischemia, for instance, oxidative metabolism comes to a halt. Even though anaerobic glycolysis increases in a compensatory fashion, this process is unable to supply sufficient amounts of ATP to maintain long-term viability and function. As a consequence, concentrations of ATP may fall by more than 70% within 20 minutes after the onset of ischemia [151]. Intracellular Ca^{2+} overloading, including cytosolic and mitochondrial, occurs as a pathophysiological response integral to the induction of oxidative stress and the subsequent appearance of cell injury. An altered redox state, where ROS and

RNS overwhelm endogenous antioxidant defenses, leads to a concentration-dependent loss of cardiomyocytes. Investigations from various groups have so far revealed that zinc can inhibit complex I, but does not increase ROS production and restores ATP generation after ischemia/reperfusion.

Ca^{2+}- Overload and Zn Homeostasis

Intracellular Ca^{2+} dramatically increases during ischemia and subsequent reperfusion, in part due to the activation of Na^{+}/Ca^{2+} exchanger and opening of L-type Ca^{2+}-channels, in conjunction with inhibition of sarcolemmal and sarcoplasmic reticular Ca^{2+}-pumps during hypoxia and activation of Na^{+}/H^{+}-exchanger during reperfusion. Extrusion of H^{+} initiates a net influx of Na^{+} into the cardiomyocyte leading to further increase in Na^{+}/Ca^{2+}-exchanger [152]. Reperfusion induces ATP-dependent hypercontracture of myocytes, due to Ca^{2+}-overloading leading to sarcolemmal rupture. Blockade of the L-type Ca^{2+}-channels can prevent overload and limit irreversible myocardial damage, provided these agents are administered prior to ischemia [153,154]. $MgSO_4$ possesses Ca^{2+}-channel blocking properties and prevents Ca^{2+}-overload during reperfusion [155]. Conversely, intracellular Mg^{2+} levels are depressed during acute myocardial infarction [156], while supplemental Mg^{2+} has been reported to reduce infarct size when administered prior to reperfusion by reducing Ca^{2+} - overloading [157,158]. In earlier studies, Zn^{2+} inhibited ryanodine binding to sarcoplasmic reticulum vesicles in cardiac muscle and has an inhibitory effect in intracellular Ca^{2+} release mechanisms in smooth muscle, skeletal muscle and cardiomyocytes [159-161] and thus has a potential to block intracellular Ca^{2+} loading. Iterations in Ca^{2+} balance, however, rarely occur in isolation. The translocation of Zn^{2+} to tissues, where it serves as an anti-oxidant, is intrinsically coupled to intracellular Ca^{2+} overloading that acts as a pro-oxidant. The relative levels of pro-oxidant: anti-oxidant determines the heart's redox state and fate of myocytes. Intracellular cation shifts, particularly Ca^{2+} and Zn^{2+}, result in excessive accumulation of Ca^{2+}, converge on mitochondria to induce oxidative stress and raise the opening potential of their inner membrane permeability transition pore (mPTP). The ensuing loss of intracellular cationic homeostasis and diminished adenosine triphosphate (ATP) production, together with osmotic swelling of mitochondria, lead to cell death and organellar degeneration. Thus, intracellular Ca^{2+} overloading is intrinsically coupled to Zn^{2+} entry through L-type Ca^{2+} channels. Extracellular zinc exposure at micromolar concentrations induces a reversible reduction

(30%) of I_{Ca} with no change in current-voltage relationship. On the other hand, an increase of intracellular free-zinc concentration, $[Zn^{2+}]_i$, from normal (< 1 nM) to approximately 7 nM with application of 10 μM Zn-pyrithione exposure causes an inhibition of 33 ± 6% in the peak of the I_{Ca} and alters the voltage dependency of L-type Ca^{2+} channels with a 10-mV left shift and a hump around -40 mV in its current-voltage relation [162]. Mitochondrial dysfunction affects cell viability through a wide array of events. Loss of ATP synthesis and increased ATP hydrolysis, impairment in ionic homeostasis of Ca^{2+}, Zn^{2+}, formation of reactive oxygen species and the release of pro-apoptotic proteins are all recognized as key factors in the generation of irreversible damage [163,164,135]. Further, an increase in cytosolic free $[Zn^{2+}]_i$ serves as an antioxidant – a negative feedback response to $[Ca^{2+}]_i$ overloading. Increased $[Zn^{2+}]_i$ activates a Zn-finger metal-response element transcription factor (MTF-1), which is central to the transcriptional expression of such antioxidant defenses as MT-1, Cu,Zn-superoxide dismutase (SOD) and glutathione synthase [165]. This capacity of zinc to counterbalance the pro-oxidant $[Ca^{2+}]_i$ may be limited, given the prevailing Zn^{2+} losses during I/R. These findings underscore the importance of coupled dyshomeostasis of Ca and Zn that accompanies I/R and clearly indicate that Zn^{2+} ionophores will be useful in modulating intracellular Zn^{2+} concentrations and has the capacity to modulate Ca^{2+} current and prevent Ca^{2+} overloading that is implicated in myocardial ischemia reperfusion injury.

Cellular Metabolic Stress in I/R and Modulation of Growth Factor Receptor by Zinc

All the events mentioned above represent a form of metabolic stress and, in some cases, oxidative stress, which is known to produce protein denaturation within the cell. Interestingly, an increase in denatured or malformed proteins in the cell have been reported to increase the synthesis of the so-called stress or heat shock proteins [166]. Hsp70 is induced to high levels after conditions similar to those encountered during myocardial ischemia. Ligation of the left anterior descending coronary artery in the heart of experimental animals for several hours produces acute myocardial ischemia, which was found to induce the synthesis of Hsp70 [167,168]. Using a cultured rat neonatal cardiomyocyte model simulating hypoxia/reoxygenation, it was observed that ATP depletion and cellular injury in the form of progressive release of arachidonic acid into the medium correlated well, and the expression of Hsp70 mRNA preceded these processes [169]. Induction of these stress proteins and molecular chaperones appears critical during this dynamic shift from steady state of tissue growth to periods of metabolic stress, to coordinate the stability and activation of a large sets of intracellular proteins [170-172]. To accomplish this task, the chaperone Hsp90, and additional proteins termed co-chaperones form the dynamic complex known as the Hsp90 chaperone machine. Many cells use the chaperone machinery to protect an array of mutated or modified proteins from misfolding and degradation. Therefore, HSp90 is recognized as a crucial mediator of cellular homeostasis, a function that it accomplishes partly by facilitating numerous transient low-affinity protein-protein interactions [173-175]. Hsp90 is ATP-dependent and the ATPase site of the chaperone is responsible for the autophosphorylation required to enable Hsp90 to interact with client proteins [176,177]. Some of the client proteins are involved in tyrosine kinase signaling (IGFR1, ErbB2, VEGFR2, c-src), the mitogen activated protein kinase (MAPK) pathway (raf-1, B-raf, MEK), the PI3K pathway (PDK1, Akt), eNOS and NF-kB signaling (the IkB kinases). These proteins vary in the degree to which Hsp90 inhibition reduces their half-lives and activities. ErbB2 is one of the client proteins that is involved in heart development and downstream signaling in myocardial ischemia/reperfusion.

The epidermal growth factor receptor (EGFR) is the prototypical member of the ErbB family that mediates multiple cellular responses in physiologic processes as well as pathophysiological states [178]. It has been well documented that EGFR can be directly activated or transactivated in response to variety of stimuli [179]. In the case of direct activation, the EGFR is regulated by binding of various polypeptide ligands that contain a 6-kDa domain homologous to EGF such as heparin-binding EGF (HB-EGF), transforming growth factor -α (TGF-α), amphiregulin, betacellulin, and epiregulin [180,181]. Following ligand binding, the EGFR undergoes homo- or heterodimerisation and subsequent auto-phosphorylation of multiple tyrosine residues in the C-terminal cytoplasmic portion of the molecule that serve as binding sites for cytosolic signaling proteins containing Src homology 2 domains and phosphotyrosine-binding domains [182]. Several in vivo autophosphorylation sites have been identified in the EGFR: three major (Tyr 1068, 1148, 1173) and two minor (Tyr 992 and 1086) [183,184]. The binding of downstream signaling proteins to the phosphorylated EGFR initiates multiple signaling cascades which culminate in pleiotropic biological responses such as mitogenesis, enhanced cell motility, protein secretion and differentiation [185]. In addition to its cognate ligands, the EGFR is "transactivated" by physioilogic stimuli such as G protein-coupled receptor (GPCR) ligands, cytokines, and chemokines, or nonphysiologic agents including arsenite, ultraviolet radiation, heat shock, and oxidants [178]. The Her-2/neu oncogenes, also known as ErbB2 is a transmembrane receptor tyrosine kinase that belongs to EGFR [186,187]. The importance of ErbB2 in normal cardiac development and physiology was demonstrated in mice by cardiac-specific knock-out of ErbB2 [188,189]. The mice were initially normal, but developed cardiomyopathy as adults and showed increased DNA fragmentation [188]. Recently, Grazette et al. studied the effects of ErbB2 blockade on cardiomyocyte survival, and showed that ErbB2 antibody caused a loss of mitochondrial membrane potential and an increase in cell death. Primary cultures of neonatal rat cardiac myocytes exposed to anti-ErbB2 antibody (anti-ErbB2 Ab) for up to 24 hours decreases cell viability, affects mitochondrial function. These effects with ErbB2 inhibition have been related to a dramatic increase in the expression of Bcl-2 family protein, Bcl-xS, and decreased levels of Bcl-xL. Also, there was a time dependent increase in mitochondrial translocation and oligomerization of Bax, followed by cytochrome c release and caspase activation. These events were associated with loss of mitochondrial membrane potential and decrease in ATP levels [190]. The mechanism for the deleterious effects of ErbB2 blockade remains unclear, but a recent report showed that activation of ErbB2 reduces doxorubicin-induced oxidative stress in cardiomyocytes [191]. Yet another study showed that ErbB2 levels decreased in an animal model of myocardial ischemia and in patients with ischemic cardiomyopathy [192]. Previous reports also suggested that PI3K and Akt have a role in ErbB2-mediated signal transduction [192,193]. Further evidence accumulated from recent studies suggests that exposure to exogenous zinc ions may trigger EGFR signaling in cardiomyocytes through a trans-activation mechanism involving c-Src kinase activity [194,195]. In vitro studies have shown that ErbB2 does not directly mediate anti-apoptotic effecs, although there are reports suggesting that ErbB2 can directly couple to PI3K [196]. Activation of Akt through PI3K is well known to couple to receptor tyrosine kinases, including those in ErbB family, and this pathway has been shown to inhibit apoptosis through inactivation of caspases [197]. In primary neonatal rat cardiac myocytes subjected to hypoxia/reoxygenation, zinc pyrithione treatment at reperfusion significantly attenuated I/R-induced decline in ErbB2 levels. Treatment with geldanamycin, an inhibitor of Hsp90 interactions with client proteins, induced the release of ErbB2 and eliminated the protective effect of zinc pyrithione. Similarly, incubation of myocytes with an anti-ErbB2 antibody (B10) or wortmannin, an inhibitor of PI3K abrogated the protective effect of zinc, providing evidence in favor of zinc pyrithione to modulate the expression and interaction of ErbB2, Hsp90 and PI3K/Akt. Thus, loss of ErbB2 function, whether through antibody blockade or enhanced degradation due to Hsp90 inhibitor, geldanamycin dramatically decreased the cytoprotective effect of zinc, suggesting that activity of Hsp90 is essential for maintaining the functional stability of ErbB2, Akt in myocytes subjected to I/R [144]. However, the signaling pathways that relay Zn^{2+} stimulation to EGFR or ErbB2 activation in cardiomyocytes during ischemia/reperfusion are still not fully understood. It is well-established that the receptor tyrosine kinase ErbB2 has a long half-life at the plasma membrane, where it appears to be the preferred heterodimerisation partner for the other ErbB family members [198-202]. Inhibition of the chaperone HSP-90 with geldanamycin leads to downregulation of many HSP-90 client proteins, of which ErbB2 is among the most affected [203]. HSP-90 inhibition causes ErbB2 to become ubiquitinylated by the co-chaperone CHIP [204,205], however the subsequent events have remained elusive and several different cellular mechanism have been suggested to take part in the downregulation of ErbB2. Assuming that ErbB2 is a recycling receptor [206,207], geldanamycin could cause a shift in the traffic of ErbB2 from a recycling pathway to a degradative, lysosomal pathway as suggested by Austin and co-workers [206]. However, if ErbB2 is an internalization-resistant receptor [208-210], then it is more plausible that geldanamycin stimulates the internalization of ErbB2 from the plasma membrane, resulting in delivery to lysosomes. Additionally, Tikhomirov and Carpenter observed that ErbB2 becomes fragmented after geldanamycin stimulation and they proposed a model involving cleavage of ErbB2 kinase domain by cytoplasmic proteases, followed by lysosomal and proteasomal degradation of the N-terminal and C-terminal parts of ErbB2, respectively [211]. Additionally, proteasomal inhibitors can efficiently rescue ErbB2 from geldanamycin-induced downregulation [212], and this had led to a wide acceptance of proteasomal protelysis as the mechanism degrading ErbB2 [213-217]. Following the notion that ErbB2 expression, stability and signaling is critical for cardioprotective effect, it is indeed clear now that zinc pyrithione can modulate the I/R-induced degradation of ErbB2 and could be a novel and powerful therapeutic approach for I/R.

It is well known that the proteasome system regulates the proteolysis of proteins involved in a variety of cellular processes including apoptosis, differentiation, inflammation and proliferation. Proteasome inhibition by clasto-lactacystin β-lactone in cultured neonatal rat cardiomyocytes has been shown to prevent cardiomyocytes necrosis and apoptosis after exposure to anoxia/reoxygenation [218]. Degradation of the endogenous apoptosis inhibitor ARC was mediated by the UPS [219]. ARC was markedly decreased in the mouse heart following I/R, suggesting that proteasome inhibition at the onset of reperfusion may block ARC degradation and protect ischemic myocardium against reperfusion-induced apoptosis. In addition, proteasome inhibition has also been found to induce heat shock proteins in cardiomyocytes and exert cardioprotection after oxidative stress [220,221]. It is interesting to learn that zinc pyrrolidine dithiocarbamate was shown to inhibit the ubiquitin-proteasome pathway and result in the accumulation of several proteasomal substrates with extended half-lives [222]. Intriguingly, zinc also modulates activities of several transcription factors including NF-kB and AP-1 that are implicated in ischemic injury [97,223]. The effect of proteasome inhibition during myocardial ischemia on cardiac function have been controversial, as both beneficial and deleterious effects have been reported [224]. Nonetheless, recent data from both in vitro and in vivo studies support the concept that short-term (acute) proteasome inhibition under certain conditions may serve as a novel therapeutic strategy for myocardial ischemia and reperfusion, although the therapeutic potential of proteasome inhibition in human cardiac disease is yet to be tested.

Therefore, the literature accumulating so far suggests that the stability of ErbB2 requires its association with Hsp90 [216]. Inhibitors of Hsp90 induce the release of ErbB2 resulting in subsequent proteasomal degradation [175]. The chaperone activity of Hsp90 requires ATP, which binds with approximately 10-fold lower affinity than ADP. Lowering ATP concentrations in isolated myocytes triggers rapid dissociation of Hsp90 from ErbB2 and degradation of ErbB2 along with other client protein). Diminished expression of ErbB2 has been observed in both animal models of ischemic heart disease and in human ischemic cardiomyopathy. Recent experimental evidence indicates that one of the downstream growth signaling effectors of ErbB2 is PI3 kinase. With zinc already known for its ability to activate PI3 kinase/Akt and targeting the mitochondrial permeability transition pore (mPTP) via glycogen synthase kinase 3β (GSK-3 β) during myocardial I/R injury, it has now emerged that zinc pyrithione can modulate ErbB2 expression during myocardial I/R and this is indeed responsible for the downstream signaling events such as Akt phosphorylation, reduced activation of caspase-3and prevention of apoptosis. This protective effect of zinc pyrithione through increased expression of ErbB2 is abolished by pretreatment with Hsp90 inhibitor, geldanamycin or PI3 kinase inhibitor, wortmannin, suggesting the critical link between these three proteins in mediating the cardioprotection.

Thus, zinc seems to be critical in regulating the stability of ErbB2, both by maintaining the ATP levels and also by facilitating the downstream activation of PI3K, Akt. It would be worth testing if zinc is required for functional stability of the Hsp90-ErbB2 complex during I/R.

The protection afforded by zinc and its ionophores suggests that it can prevent or attenuate the rise in intracellular calcium, ROS and loss of mitochondrial membrane potential during reperfusion. As often seen with many emerging drugs in reperfusion injury, zinc also appears to target several pathways, such as MAPK (ERK ½), PI3K, Akt, adenosine receptor, Protein kinase C, G and A, mitochondrial complex I, NF-kB, ubiquitin proteasome system, ErbB2 and NADPH oxidase. Although the data is convincing from experimental studies, it is important to assess protection after longer, more clinically relevant periods of ischemia. It is also imperative that treatment be initiated before or immediately at the start of reperfusion, for example, with cardiac surgery or angioplasty. It is worth considering the zinc ionophore pyrithione for further studies, given the multiple targets and mechanisms of preventing cell death and improving cardiac function during I/R.

REFERENCES

[1] M. Hambidge, *J. Nutr.* 130, 1344 (2000).

[2] W. Maret and A. Krezel, *Mol. Med.* 13, 371 (2007).

[3] D.S. Auld, *Biometals* 14,271 (2001).

[4] D. Atar, P.H. Backx, M.M. Appel, W.D. Gao and E. Marban, *J. Biol. Chem.* 270, 2473 (1995).

[5] M. Ayaz and B. Turan, Am. *J. Physiol. Heart Circ. Physiol.* 290,1071 (2006).

[6] J.S. Becker, U. Breuer, H.F. Hsieh, T. Osterholt, U. Kumbtabtim , B. Wu, A. Matusch, J.A. Caruso and Z. Qin, *Anal. Chem.* 82, 9528 (2010).

[7] P.D. Zalewski, A.Q. Truong-Tran, D. Grosser, L. Jayaram, C. Murgia and R.E. Ruffin, *Pharmacol. Ther.*105, 127 (2005).

[8] R.A. Colvin, N. Davis, R.W. Nipper and P.A. Carter, *Neurochem. Int.* 36, 539 (2000).

[9] R.A. Colvin, *Neuroreport*, 9,3091 (1998)

[10] A.H. Kim, C.T. Sheline, M. Tian, T. Higashi, R.J. McMahon, R.J. Cousins and D.W. Choi, *Brain Res.* 886, 99 (2000).

[11] E. Ohana, D. Segal, R. Palty, D. Ton-That, A. Moran, S.L. Sensi, J.H. Weiss, M. Hershfinkel and I. Sekler, *J. Biol. Chem.* 279, 4278 (2004).

[12] B.M. Palmer et al., *J. Struct. Biol.* 155, 12 (2006).

[13] L.C. Costello and R.B. Franklin, *Prostate*, 35, 285 (1998).

[14] W.E.C. Wacker, D.D. Ulmer, B.L. Vallee, *New Engl. J. Med.* 255, 449 (1956).

[15] J.A. Halsted and J.C. Smith Jr.,*Lancet*, 1, 322 (1970).

[16] R.D. Lindeman, A.A. Yunice, D.J. Baxter, L.R. Miller and J. Nordquist, *J. Lab. Clin. Med.* 81, 194 (1973).

[17] A.M. Handjani, J.C. Smith Jr., J.B. Herrmann and J.A. Halsted, *Chest*, 65, 185 (1974).

[18] O. Oster, M. Dahm and H. Oelert. *Eur. Heart J.* 14, 770 (1993).

[19] M.Ahmad, Salahuddin, B.M. Mathew, S. Kumar, T.D. Seth, M.Z. Hasan and S.Q. Mahdi, *Adv. Myocardiol.* 2, 171 (1980).

[20] B.M. Mathew, S. Kumar, M.S. Ahmad, T.D. Seth, M.Z. Hasan and S.A. Jamil, *Jpn. Circ. J.* 42, 353 (1978).

[21] F. Kosar, I. Sahin, N. Acikgoz et al., *Biol. Trace Elem. Res.* 107, 1 (2005).

[22] S. Ripa, R. Sipa and S. Giustiniani, *Minerva Med.* 89, 397 (1998).

[23] O. Oster, *Clin. Chim. Acta.* 214, 209 (1993).

[24] G. Topuzoglu, A.R. Erbay, A.B. Karul et al., *Biol. Trace Elem. Res.* 95, 11 (2003).

[25] M.D. Apostolova, I.A. Ivanova and M.G. Cherian, *Toxicol. Appl. Pharmacol.* 159, 175 (1999).

[26] M.H. Kown, T. Van der Steenhoven, F.G. Blankenberg, G. Hoyt, G.J. Berry, J.F. Tait, H.W. Strauss and R.C. Robbins, *Circulation,* 102, 228 (2000).

[27] S.R. Powell, L. Aiuto, D. Hall and A.J. Tortolani, *J. Thorac. Cardiovasc. Surg.* 110, 1642 (1995).

[28] H.I. Afridi, T.G. Kazi, M.K. Jamali and G.Q. Shar. *Biol. Trace Elem. Res.* 113, 19 (2006).

[29] L.H. Opie, *Circulation*, 80, 1049 (1989).

[30] D.J. Hearse, *Cardiovasc. Res.* 28, 1737 (1996).

[31] G. Karagulova, Y. Yue, A. Moreyra, M. Boutjdir and I. Korichneva, *J. Pharmacol. Exp. Ther.* 321, 517 (2007).

[32] H.M. Piper, K. Meuter and C. Schafer, *Ann.Thorac. Surg.* 75, 644 (2003).

[33] L.M. Buja, *Cardiovasc. Pathol.* 14, 170 (2005).

[34] R.A. Kloner and S.H. Rezkalla. *J. Am. Coll.Cardiol.* 44, 276 (2004).

[35] R. Bolli, L. Becker, G. Gross et al., *Circ. Res.* 95, 125 (2004).

[36] B. Turan, H. Fliss and M. Desilets, *Am. J. Physiol. Heart Circ. Physiol.* 272, 2095 (1997).

[37] S. Yamasaki, K. Sakata-Sogawa, A. Hasegawa, T. Suzuki, K. Kabu, E. Sato et al., *J. Cell Biol.* 177, 637 (2007).

[38] R.C. Thomas, J.A. Coles and J.W. Deitmer, *Nature*, 350, 464 (1991).

[39] D.J. Eide, *Biochim. Biophys. Acta.* 1763, 711 (2006).

[40] R.J. Cousins, J.P. Liuzzi and L.A. Lichten, *J. Biol. Chem.* 281, 24085 (2006).

[41] D.J. Eide, *Pflugers Arch.* 447, 796 (2004).

[42] R.D. Palmiter and L. Huang, *Pflugers Arch.* 447, 744 (2004).

[43] J.P. Liuzzi and R.J. Cousins, *Annu. Rev. Nutr.* 24, 151 (2004).

[44] K. Ishihara, T. Yamazaki, Y. Ishida, T. Suzuki, K. Oda et al., *J. Biol. Chem.* 281, 17743 (2006).

[45] W. Maret, *J. Nutr.* 133, 1460 (2003).

[46] W. Maret, *J. Nutr.* 130, 1455 (2000).

[47] A. Krezel and W. Maret, *J. Am. Chem. Soc.* 129, 10911 (2007).

[48] W. Maret, Proc. Natl. Acad. Sci. USA. 91, 237 (1994).

[49] Y.J. Kang, G. Li and J.T. Saari, *Am. J. Physiol. Heart Circ. Physiol.* 276, 993 (1999).

[50] G.W. Wang, Z. Zhou, J.B. Klein and J.Y. Kang, *Am. J. Physiol. Heart Circ. Physiol.* 280, 2292 (2001).

[51] D.U. Spahl, D. Berendji-Grun, C.V. Suschek, V. Kolb-Bachoffen and K.D. Kroncke, *Proc. Natl. Acad. Sci. USA.* 100, 13952 (2003).

[52] I. Sekler, S.L. Sensi, M. Hershfinkel and W.F. Silverman, *Mol .Med.* 13, 337 (2007).

[53] S.L. Sensi, H.Z. Yin and J.H. Weiss, Eur. J. Neurosci. 12, 3813 (2000).

[54] S.L. Sensi, D. Ton-That and J.H. Weiss. *Neurobiol. Dis.* 10, 100 (2002).

[55] V.P. Skulachev, V.V. Chistyakov, A.A. Jasaitis and E.G. Smirnova, *Biochem. Biophys. Res. Commun.* 26, 1 (1967).

[56] D. Kleiner, *Arch. Biochem. Biophys.* 165, 121 (1974).

[57] T.A. Link and G. von Jagow, *J. Biol. Chem.* 270, 25001 (1995).

[58] J. Wudarczyk, G. Debska and E. Lenartowicz, Arch. Biochem. Biophys. 363, 1 (1999).

[59] A.M. Brown, B.S. Kristal, M.S. Effron, A. I. Shestopalov, P.A. Ullucci, K.F. Sheu, J.P. Blass and A.J. Cooper, *J. Biol. Chem.* 275, 13441 (2000).

[60] D. Jiang, P.G. Sullivan, S.L. Sensi, O. Steward and J.H. Weiss, *J. Biol. Chem.* 276, 47524 (2001).

[61] B. Ye, W. Maret and B.L. Vallee, *Proc. Natl. Acad. Sci. USA.* 98, 2317 (2001).

[62] I.G. Gazaryan, B.F. Krasnikov, G.A. Ashby, R.N. Thorneley, B.S. Kristal and A.M. Brown, *J. Biol. Chem.* 277, 10064 (2002).

[63] S.R. Powell, D. Hall, L. Aiuto, R.A. Wapnir, S. Teichberg and A. Tortolani, *Am. J. Physiol. Heart Circ. Physiol.* 266, 2497 (1994).

[64] H. Tong, H.A. Rockman, W.J. Koch, C. Steenbergen and E.G. Murphy, *Circ. Res.* 94, 1133 (2004).

[65] G. Chanoit, S. Lee, J. Xi, M. Zhu, R.A. McIntosh, R.A. Mueller et al., *Am. J. Physiol. Heart Circ. Physiol.* 295, 1227 (2008).

[66] S. Lee, G. Chanoit, R.A. McIntosh, D.A. Zvara and Z. Xu, *Am. J. Physiol. Heart Circ. Physiol.* 297, 569 (2009).

[67] Y. Jang, H. Wang, J. Xi, R.A. Mueller, E.A. Norfleet and Z. Xu, *Cardiovasc. Res.* 75, 426 (2007).

[68] P. Ping, H. Takano, J. Zhang, X.L. Tang, Y. Qiu, R.C. Li, S. Banerjee, B. Dawn, Z. Balafonova and R. Bolli, *Circ. Res.* 84, 587 (1999).

[69] H. Tong, W. Chen, C. Steenbergen and E. Murphy, *Circ. Res.* 87, 309 (2000).

[70] A. Barthel, E.A. Ostrakhovitch, P.L. Walter, A. Kampkotter, L.O. Klotz, *Arch. Biochem. Biophys.* 463, 177 (2007).

[71] M.M. Mocanu and D.M. Yellon, *Br. J. Pharmacol.* 150, 833 (2007).

[72] W. Wu, X. Wang, W. Zhang, W. Reed, J.M. Samet, Y.E. Whang and A.J. Ghio, *J. Biol. Chem.* 278, 28258 (2003).

[73] M.M. Mocanu and D.M. Yellon, *Am. J. Physiol. Heart Circ. Physiol.* 296, 233 (2009).

[74] T.A. Fralix, E. Murphy, R.E. London and C.Steenbergen, *Am. J. Physiol. Cell Physiol.* 264, 986 (1993).

[75] Z. Xu, S.S. Park, R.A. Mueller, R.C. Bagnell, C. Patterson and P.G. Boysen, *Cardiovasc. Res.* 65, 803 (2005).

[76] R. McIntosh, S. Lee, A.J. Ghio, J. Xi, M. Zhu, X. Shen, G. Chanoit, D.A. Zyara and Z. Xu, *J. Mol. Cell. Cardiol.* 49, 41 (2010).

[77] C.E. Murray, R.B. Jennings and K.A. Reimer, *Circulation*, 74, 1124 (1986).

[78] G.P. Matherne et al., *Pediatr. Res.* 28, 348 (1990).

[79] T.A. Dorheim et al., *J. Surg. Res.* 48, 491 (1990).

[80] J.R. De et al., *Eur. J. Pharmacol.* 441, 165 (2002).

[81] T. Miura et al., *J. Am. Coll. Cardiol.* 35, 238 (2000).

[82] Z. Xu et al., *J. Mol. Cell. Cardiol.* 32, 2339 (2000).

[83] Z. Xu, J.M. Downey and M.V. Cohen, *Heart Dis.* 5, 368 (2003).

[84] G.J. Smits et al., *J. Pharmacol. Exp. Ther.* 286, 611 (1998).

[85] M.S. Sharpley and J. Hirst, *J. Biol. Chem.* 281, 34803 (2006).

[86] N.C. Moss, W.E. Stansfield, M.S. Willis, R.H. Tang and C.H. Selzman, *Am. J. Physiol. Heart Circ. Physiol.* 293, 2248 (2007).
[87] N.C. Moss, R.H. Tang, M. Willis, W.E. Stansfield, A.S. Baldwin and C.H. Selzman, *J. Thorac. Cardiovasc. Surg.* 136, 1274 (2008).
[88] B. Campbell, J. Adams, Y.K. Shin and A.M. Lefer, *J. Mol. Cell. Cardiol.* 31, 467 (1999).
[89] J. Pye, F. Ardeshirpour, A. McCain, D.A. Bellinger, E. Merrick, J. Adams et al., *Am. J. Physiol. Heart Circ. Physiol.* 284, 919 (2003).
[90] W.E. Stansfield, N.C. Moss, M.S. Willis, R. Tang and C.H. Selzman, *Ann. Thorac. Surg.* 84, 120 (2007).
[91] M.R. Hoffmeyer, R. Scalia, C.R. Ross, S.P. Jones and D.J. Lefer, *Am. J. Physiol. Heart Circ. Physiol.* 279, 2824 (2000).
[92] U. Zabel, R. Schreck and P.A. Baeuerle, *J. Biol. Chem.* 266, 252 (1991).
[93] J.P. Yang, J.P. Merin, T. Nakana, T. Kato, Y. Kitade and T. Okamoto, *FEBS Lett.* 361, 89 (1995).
[94] J.A. Shumilla, K.E. Wetterhahn and A. Barcowsky, *Arch. Biochem. Biophys.* 349, 356 (1998).
[95] P. Connel, V.M. Young, M. Toborek, D.A. Cohen, S. Barve, C.J. McCain and B. Hennig, *J. Am.Coll. Nutr.* 16, 411 (1997).
[96] C.H. Kim, J.H. Kim, J. Xu, C.Y. Hsu and Y.S. Ahn, *J. Neurochem.* 72, 1586 (1999).
[97] C.H. Kim, J.H. Kim, C.Y. Hsu and Y.S. Ahn, *FEBS Lett.* 449, 28 (1999).
[98] R. Schrek, B. Meier, D.N. Mannel, W. Droge and P.A. Baeuerle, *J. Exp. Med.* 175, 1181 (1992).
[99] R. Schreck, B. Meier, D.N. Mannel, W. Droge and P.A. Baeuerle, *Free Rad. Res. Commun.* 17, 221 (1992).
[100] R. Bessho, K. Matsubara, M. Kubota, K. Kuwkado, H. Hirota, Y. Wakazono, Y.W. Lin, A. Okuda, M. Kawi, R. Nishikomori and T. Heike, *Biochem. Pharmacol.* 48, 1883 (1994).
[101] P.D. Zalewski, I.J. Forbes and W.H. Betts, *Biochem. J.* 296, 403 (1993).
[102] C.H. Kim, J.H. Kim, S.J. Moon, K.C. Chung, C.Y. Hsu, J.T. Seo, Y.S. Ahn, *Biochem. Biophys. Res. Commun.* 259, 505 (1999).
[103] C.H. Kim, J.H. Kim, J. Lee, C.Y. Hsu and Y.S. Ahn, *J. Biol. Chem.* 384, 143 (2003).
[104] M.S. Hayden and S. Ghosh, *Cell.* 132, 344 (2008).
[105] S. Ghosh, M.J. May and E.B. Kopp, *Annu. Rev. Immunol.* 16, 225 (1998).
[106] P.A. Baeuerle and D. Baltimore, *Science*, 242, 540 (1988).

[107] S. Frantz, J. Tillmanns, P.J. Kuhlencordt, I. Schmidt, A. Adamek, C. Dienesch, T. Thum, S. Gerondakis, G. Ertl and J. Bauersachs, *Am. J. Path.* 171, 507 (2007).
[108] B. Chandrasekar and G.L. Freeman, *FEBS Lett.* 401, 30 (1997).
[109] N. Shimizu, M. Yoshiyama, T. Omura, A. Hanatani, S. Kim, K. Takeuchi, H. Iwao and J. Yoshikawa, *Cardiovasc. Res.* 38, 116 (1998).
[110] S. Mustapha, A. Kirshner, D. De Moissac and L.A. Kirschenbaum, *Am. J. Physiol. Heart Circ. Physiol.* 279, 939 (2000).
[111] N. Maekawa, H. Wada, T. Kanda, T. Niwa, Y. Yamada, K. Saito, H. Fujiwara, K. Sekikawa and M. Seishima, *J. Am. Coll. Cardiol.* 39, 1229 (2002).
[112] A. Misra, S.B. Haudek, P. Knuefermann, J.G. Vallejo, Z.J. Chen, L.H. Michael, N. Sivasubramanian, E.N. Olson, M.L. Entman and D.L. Mann, *Circulation*, 108, 3075 (2003).
[113] T. Hamid, S.Z. Guo, J.R. Kingery, X. Xiang, B. Dawn and S.D. Prabhu, *Cardiovasc. Res.* 89, 129 (2011).
[114] J.D. Lambeth, *Nat. Rev. Immunol.* 4, 181 (2004).
[115] B.M. Babior, *Blood*, 93, 1464 (1999).
[116] L. Xiao, D.R. Pimentel, J. Wang, K. Singh, W.S. Colucci and D.B. Sawyer, *Am. J. Physiol. Cell. Physiol.* 282, 926 (2002).
[117] K.K. Griendling, C.A. Minieri, J.D. Ollerenshaw and R.W. Alexander, *Circ. Res.* 74, 1141 (1994).
[118] A. Gorlach, R.P. Brandes, K. Nguyen, M. Amidi, F. Dehghani and R. Busse, *Circ. Res.* 87, 26 (2000).
[119] T. Ago, T. Kitazono, H. Ooboshi, T. Iyama, Y.H. Han, J. Takada et al., *Circulation*, 109, 227 (2004).
[120] J.K. Bendall, A.C. Cave, C. Heymes, N. Gall and A.M. Shah, *Circulation*, 105, 293 (2002).
[121] J.A. Byrne, D.J. Grieve, J.K.Bendall, J.M. Li, C. Gove, J.D. Lambeth et al., *Circ. Res.* 93, 802 (2003).
[122] I. Cucoranu, R. Clempus, A. Dikaloya, P.J. Pehlan, S.Ariyan, S.Dikaloy and D> Sorescu, *Circ. Res.* 97, 900 (2005).
[123] J.M. Li and A.M. Shah, *J. Biol. Chem.* 277, 19952 (2002).
[124] T. Fukui, M. Yoshiyama, A. Hantani, T. Omura, J. Yoshikawa and Y. Abe, *Biochem. Biophys. Res. Commun.* 281, 1200 (2001).
[125] H. Cai, K.K. Griendling and D.G. Harrison, *Trends Pharmacol. Sci.* 24, 471 (2003).
[126] D.I. Brown and K.K. Griendling, *Free Rad. Biol. Med.* 47, 1239 (2009).

[127] M.H. Bai, W. Dai, Y.J. Li and C.P. Hu, *Can. J. Physiol. Pharmacol.* 89, 177 (2011).
[128] V. Kasi, S. Bodiga, U.N. Kommuguri, S. Sankuru and V.L. Bodiga, *Biochem. Biophys. Res. Commun.* 410, 270 (2011).
[129] J. Duan and M. Karmazyn, *Can. J. Physiol. Pharmacol.* 67, 704 (1989).
[130] W. Flameng, J. Andres, P. Ferdinande, M. Mattheussen and H. Van Belle, *J. Mol. Cell. Cardiol.* 23, 1 (1999).
[131] E.J. Lesnefsky, B. Tandler, J. Ye, T.J. Slabe, J. Turkaly and C.L. Hoppel, *Am. J. Physiol. Heart Circ. Physiol.* 273, 1544 (1997).
[132] W. Rouslin, Am. J. Physiol. Heart Circ. Physiol. 244, 743 (1983).
[133] G.K. Asimakis and V.R. Conti, *J. Mol. Cell. Cardiol.* 16, 439 (1984).
[134] E.J. Lesnefsky, T.I. Gudz, C.T. Migita, M.Ikeda-Saito, M.O. Hassan, P.J. Turkaly and C.L. Hoppel, *Arch. Biochem. Biophys.* 385, 117 (2001).
[135] E.J. Lesnefsky, S. Moghaddas, B. Tandler, J. Kerner and C.L. Hoppel, *J. Mol. Cell. Cardiol.* 33, 1065 (2001).
[136] H.M. Piper, O. Sezer, M. Schleyer, P. Schwartz, J.F. Hutter and P.G. Spieckermann, *J. Mol. Cell. Cardiol.* 17, 885 (1985).
[137] H. Ueta, R. Ogura, M. Sugiyama, A. Kagiyama and G. Shin, *J. Mol. Cell. Cardiol.* 22, 893 (1990).
[138] H. Nohl, *Ann. Biol. Clin (Paris),* 52, 199 (1994).
[139] J.F. Turrens, M. Becoi, J. Barilla, U.B. Chavez and J.M. McCord, *Biochem. J.* 191, 421 (1980).
[140] V. Borutaite, A. jekabsone, R. Morkuniene and G.C. Brown, *J. Mol. Cell. Cardiol.* 35, 357 (2003).
[141] M. Chen, H. He, S.Zhan, S. Krajewski, J.C. Reed and R.A. Gottlieb, *J. Biol. Chem.* 276, 30724 (2001).
[142] R.A. Gottlieb, K.O. Burleson, R.A. Kloner, B.M. Babior and R.L. Engler, *J. Clin. Invest.* 94, 1621 (1994).
[143] D.J. Hausenloy and D.M. Yellon, *J. Mol. Cell. Cardiol.* 35, 339 (2003).
[144] K. Viswanath, S. Bodiga, V. Balogun, A. Zhang and V.L. Bodiga, *Biometals*, 24, 171 (2011).
[145] V.P. Skulachev, V.V. Chistyakov, A.A. Jasaitis and E.G.Smirnova, *Biochem. Biophys. Res. Commun.* 26, 1 (1967).
[146] P. Nicholls and A.P. Singh, *Life Sci. Adv. (Biochemistry, Delhi)* 7, 321 (1988).
[147] A. Aagaard and P. Brezezinski, *FEBS Lett.* 494, 157 (2001).
[148] M.S. Sharpley and J. Hirst. *J. Biol. Chem.* 281, 34803 (2006).
[149] E. Barth, G. Stammler, B. Speiser and J. Schaper, *J. Mol. Cell. Cardiol.* 24, 669 (1992).

[150] B. O'Rourke, S. Cortassa and M.A. Aon, *Physiology (Bethesda)*, 20, 303 (2005).
[151] E. Murphy and C. Steenbergen, *Physiology (Bethesda)* 23, 115 (2008).
[152] H.M. Piper, D. Garcia-Dorado and M. Ovize, *Cardiovasc. Res.* 38, 291 (1998).
[153] D. Garcia-Dorado, P. Theroux, F. Fernandez-Aviles, J. Elizaga, J. Solares and M. Galinanes, *J. Am. Coll. Cardiol.* 10, 906 (1987).
[154] L. Higginson, A. Tang, G. Knoll and J. Calvin, *J. Am. Coll.Cardiol.* 18, 868 (1991).
[155] K.L. Woods, *Br. J. Clin. Pharmacol.* 32, 3 (1991).
[156] M.C. Haigney, B. Silver, E.Tanglao, H.S. Silverman, J.D. Hill, E. Shapiro et al., *Circulation*, 92, 2190 (1995).
[157] C.W. Christensen, M.A. Rieder, E.L. Silverstein and N.E. Gencheff, *Circulation*, 92, 2617 (1995).
[158] W.R. Herzog, M.L. Scholssberg, K.S. MacMurdy, L.R. Edenbaum, M.J. Gerber, R.A. Ogel et al., *Circulaton*, 92, 2622 (1995).
[159] X.Y. Cheng, K.Y. Chen, X. H. Zhang and P.H. Zhu, *Cell Physiol. Biochem.* 12, 119 (2002).
[160] S. Musat, X. Wang and N.S. Dhalla, *J. Cardiovasc. Pharmacol. Ther.* 3, 291 (1998).
[161] H. Wang, Q.Q. Wei, X.Y. Cheng, K.Y. Chen and P.H. Zhu, *Cell Physiol. Biochem.* 11, 83 (2001).
[162] B. Turan, *Biol. Trace Elem. Res.* 94, 49 (2003).
[163] F. Di Lisa, R. Menabo, M. Canton and V. Petronilli, *Biochim. Biophys. Acta.* 1366, 69 (1998).
[164] M.R. Duchen, *J. Physiol.* 529, 57 (2000).
[165] G.K. Andrews, *Biometals*, 14, 223 (2001).
[166] J. Ananthan, A.L. Goldberg and R. Voellmy, *Science*, 232, 522 (1986).
[167] W.H. Dillmann, H.B. Mehta, A. Barrieux, B.D. Guth, W.E. Neeley and J. Ross, *Circ. Res.* 59, 110 (1986).
[168] H.B. Mehta, B.K. Popoyich and W.H. Dillmann, *Circ. Res.* 63, 512 (1988).
[169] K. Iwai, S.H. Chi, W.H. Dillmann and R. Mestril, *Circulation*, 87, 2023 (1993).
[170] S.K. Wanglinger, K. Richter and J. Buchner, *J. Biol. Chem.* 283, 18473 (2008).
[171] R. Zhao et al., *Cell*, 120, 715 (2005).
[172] W.B. Pratt and D.O. Toft, *Exp. Biol. Med. (Maywood)*, 228, 111 (2003).

[173] W.B. Pratt, Y. Morishima and Y. Osawa, *J. Biol. Chem.* 283, 22885 (2008).

[174] M.V. Blagosklonny, *Leukemia*, 16, 455 (2002).

[175] L. Neckers, *Trends Mol. Med.* 8, 55 (2002).

[176] T.W. Schulte, S. Akinaga, T. Murakata, T. Agatsuma, S. Sugimoto, H. Nakano, Y.S. Lee, B.B.Simen, Y. Argon, S. Felts, D.O. Toft, L.M. Neckers and S.V. Sharma, *Mol. Endocrinol.* 13, 1435 (1999).

[177] T.W. Schulte, S. Akinaga, S. Soga, W. Sullivan, B. Stensgard, D. Toft and L.M. Neckers, *Cell Stress Chaperones*, 3, 100 (1998).

[178] G. Carpenter, *J. Cell. Biol.* 146, 697 (1999).

[179] A. Gschwind, E. Zwick, N. Prenzel, M. Leserer and A. Ullrich, *Oncogene*, 20, 1594 (2001).

[180] J. Schlessinger, *Cell*, 110, 669 (2002).

[181] D.S. Salomon, R. Brandt, F. Ciardiello and N. Normanno, *Crit. Rev. Oncol. Hematol.* 19, 183 (1995).

[182] T. Pawson, P. Olivier, M. Rozakis-Adcock, J. McGlade and M. Henkemeyer, *Philo. Trans. R. Soc. Lond. B. Biol. Sci.* 340, 279 (1993).

[183] J. Downward, P. Parker and M.D. Waterfield, *Nature*, 311, 483 (1984).

[184] G.M. Walton, W.S. Chen, M.G. Rosenfield and G.N. Gill, *J. Biol. Chem.* 265, 1750 (1990).

[185] J. Blenis, Proc. Natl. Acad. Sci. USA 90, 5889 (1993).

[186] L.C. Groenin, E.C. Nice and A.W. Burgess, *Growth Factors*, 11, 235 (1994).

[187] G. Lemke, *Mol. Cell. Neurosci.* 7, 247 (1996).

[188] S.A. Crone, Y.Y. Zhao, L. Fan, Y. Gu, S. Minamisawa, Y. Liu, K.L. Peterson, J. Chen, R. Kahn, G. Condorelli et al., *Proc. Natl. Acad. Sci. USA*. 99, 8880 (2002).

[189] S.A. Crone, Y.Y. Zhao, L. Fan, Y. Gu, S Minamisawa, Y. Liu, K.L. Peterson, J. Chen et al., *Nat. Med.* 8, 459 (2002).

[190] L.P. Grazette, W. Boecker, T. Matsui, M. semigran, T.L. Force, R.J. Hajjar and A. Rosenzweig, *J. Am. Coll. Cardiol.* 44, 2231 (2004).

[191] F. Imolati, D. Ott, L. Pentassuglia, M.N. Giraud, J. C. Perriard, T.M. Suter and C. Zuppinger, *J. Mol. Cell. Cardiol.* 41, 845 (2006).

[192] L.I. Gordon, M.A. Burke, A.T.K. Singh, S. Prachand, E.D. Lieberman, L. Sun, T.J. Naik, S.V. Naga Prasad and H. Ardehali, *J. Biol. Chem.* 284, 2080 (2009).

[193] S.J. Fuller, K. Sivarajah and P.H. Sudgen, *J. Mol. Cell. Cardiol.* 44, 831 (2008).

[194] W. Wu, L.M. Graves, G.N. Gill, S.J. Parsons and J.M. Samet, *J. Biol. Chem.* 277, 24252 (2002).
[195] J.M. Samet, B.J. Dewar, W. Wu and L.M. Graves, *Toxicol. Appl. Pharmacol.* 191, 86 (2003).
[196] N.J. Hellyer, M.S. Kim and J.G. Koland, *J. Biol. Chem.* 276, 42153 (2001).
[197] R.R. Baliga, D.R. Pimental, Y.Y. Zhao, W.W. Simmons, M.A. Marchionni, D.B. Sawyer and R.A. Kelly, *Am. J. Physiol. Heart Circ. Physiol.* 277, 2026 (1999).
[198] J. Baulida, M.H. Kraus, M. Alimandi, P.P. Di Fiore and G. Carpenter, *J. Biol. Chem.* 271, 5251 (1996).
[199] D. Graus-Porta, R.R. Beerli, J.M. Daly and N.E. Hynes, *EMBO J.* 16, 1647 (1997).
[200] Z. Wang, L. Zhang, T.K. Yeung and X. Chen, *Mol. Biol. Cell.* 10, 1621 (1999).
[201] Y. Yarden, *Oncology*, 61, 1 (2001).
[202] Y. Yarden and M.X. Slikowski, *Nat. Rev. Mol. Cell Biol.* 2, 127 (2001).
[203] L. Neckers and S.P. Ivy, *Curr. Opin. Oncol.* 15, 419 (2003).
[204] W. Xu, M. Marcu, X. Yuan, E. Mimnaugh, C. Patterson and L. Neckers, *Proc. Natl. Acad. Sci. USA.* 99, 12847 (2002).
[205] P. Zhou, N. Fernandes, I.L. Dodge, A.L. Reddi, N. Rao, H. Safran, T.A. DiPetrillo, D.E. Wazer, V. Band and H. Band, *J. Biol. Chem.* 278, 13829 (2003).
[206] C.D. Austin, A.M. De Maziere, P.L. Pisacane, S.M. van Dijk, C. Eigenbrot, M.X. Sliwkowski, J. Klumperman and R.H. Scheller, *Mol. Biol. Cell*, 15, 5268 (2004).
[207] Y. Yarden, *Oncology*, 61, 1 (2001).
[208] A.M. Hommelgaard, M. Lerdrup and B. van Deurs, *Mol. Biol. Cell*, 15, 1557 (2004).
[209] K.E. Longva, N.M. Pedersen, C. Haslekas, E. Stang and I.H. Madshus, *Int. J. Cancer.* 116, 359 (2005).
[210] Z. Wang, L. Zhang, T.K. Yeung and X. Chen, *Mol. Biol. Cell* 10, 1621 (1999).
[211] O. Tikhomirov and G. Carpenter, *J. Biol. Chem.* 275, 26625 (2000).
[212] E.G. Mimnaugh, C. Chavany and L. Neckers, *J. Biol. Chem.* 271, 22796 (1996).
[213] A. Citri, I. Alroy, S. Lavi, C. Rubin, W. Xu, N. Grammatikakis, C. Patterson, L. Neckers, D.W. Fry and Y. Yarden, *EMBO J.* 21, 2407 (2002).

[214] R.L. Hong, W.H. Spohn and M.C. Hung, *Clin. Cancer Res.* 5, 1884 (1999).
[215] T.D. Way, M.C. Kao and J.K. Lin, *J. Biol. Chem.* 279, 4479 (2004).
[216] W. Xu, E. Mimnaugh, M.F. Rosser, C. Nicchitta, M. Marcu, Y. Yarden and L. Neckers, *J. Biol. Chem.* 276, 3702 (2001).
[217] F.F. Zheng, S.D. Kuduk, G. Chiosis, P.N. Munster, L. Sepp-Lorenzino, S.J. Danishefsky and N. Rosen, *Cancer Res.* 60, 2090 (2000).
[218] V.E. Dosenko, V.S. Nagibin, L.V. Tumanovskaya, V.Y. Zagoriv, A.A. Moibenko and J. Vaage, *Pathophysiol.* 13, 119 (2006).
[219] Y.J. Nam, K. Mani, L. Wu, C.F. Peng, J.W. Calvert, R.S. Foo et al., *J. Biol. Chem.* 282, 5522 (2007).
[220] H. Luss, W. Schmitz and J. Neumann, *Cardiovasc. Res.* 54, 140 (2002).
[221] K. Stangl, C. Gunther, T. Frank, M. Lorenz, S. meiners, T. Ropke et al., *Biochem. Biophys. Res. Commun.* 291, 542 (2002).
[222] I. Kim, C.H. Kim, J.H. Kim, J. Lee, J.J. Choi, Z.A. Chen, M.G. Lee, K.C. Chung, C.Y. Hsu and Y.S. Ahn, *Exp. Cell Res.* 298, 229 (2004).
[223] C.H. Kim, J.H. Kim, J. Lee, C.Y. Hsu and Y.S. Ahn, *Biol. Chem.* 384, 143 (2002).
[224] X. Yu and D.C. Kem, *Cardiovasc. Res.* 85, 312 (2010).

In: Zinc: Characteristics, Uses and Benefits ISBN: 978-1-61942-523-1
Editor: Masayoshi Yamaguchi

Chapter 6

Use of Zinc Supplementation for Modulating the Immune Response

Anne Brieger and Hajo Haase*
Institute of Immunology, Medical Faculty,
RWTH Aachen University, Germany.

Abstract

The trace element zinc is ubiquitous in the human body. It has catalytic, structural or regulatory functions in more than 300 enzymes. The importance of an adequate zinc intake is marked by its impact on biological functions such as DNA replication, RNA transcription, signal transduction and redox regulation. In particular, cells of the innate and adaptive immune system depend on zinc ions, e.g., for proliferation, differentiation, and the production of cytokines. Hence, a deficiency affects the susceptibility to infections and the development of inflammation and autoimmunity. Zinc deficiency is mainly caused by malnutrition, but can also occur as a result of genetic defects of zinc transport proteins (e.g., in *Acrodermatitis enteropathica*) or secondary to diseases such as type 2 diabetes or diarrhea. Although a variety of studies have been performed regarding the therapeutic usefulness of zinc supplementation, the effects on several diseases remain unclear. Contradictory outcomes between different studies are caused by administration of diverse zinc salts and their varying bioavailability, the

* E-Mail: HHaase@ukaachen.de

problematic determination of the patients' zinc status by plasma or serum zinc level, and differences in dosage and duration of the supplementation. For example, zinc is only effective in the treatment of the common cold if given within 24h of the onset of symptoms. Furthermore, zinc supplementation was shown to be beneficial for decreasing the frequency of diarrhea and pneumonia, diseases which contribute significantly to the high infant mortality in developing countries. On the other hand, zinc excess may result in increased susceptibility to infections by inhibiting leukocyte function. This review gives an overview on populations at risk of zinc deficiency, available supplements, and the impact of zinc administration on immune cells and several diseases.

INTRODUCTION

The human body contains 2-3 g of the trace element zinc. Its homeostasis is effectively regulated by intestinal uptake and excretion [158]. The majority is found in bones and musculature, minor fractions in skin, brain, liver, and several other organs [85]. Only a very small part of zinc is circulating in the blood, mostly bound to albumin, but it is important to distribute zinc to all organs [161]. Due to the absence of a storage organ for zinc, the body depends on a constant nutritional supply to ensure sufficient zinc availability for all its biological functions [158]. More than 300 enzymes contain zinc, either involved in catalytic activity or stabilizing their structure [177]. Moreover, the structure of many other proteins also depends on zinc; especially transcription and replication factors require this metal ion, for example in zinc finger motifs [6, 177].

The immune system is one of the organs with the highest vulnerability to zinc deficiency. It depends on zinc to maintain its high rate of proliferation and differentiation, and impaired immune function as a result of limited zinc availability leads to increased morbidity and even death due to infections [150]. At the beginning of this chapter, an overview about zinc deficiency and the different nutritional supplements is given. After a short introduction to the immune system, the interactions between zinc levels and the immune response are discussed. The last part of this chapter describes the impact of zinc supplementation on several diseases, for which either a clear association between zinc metabolism and disease was found, or a beneficial impact of zinc supplementation was described.

POPULATIONS AT RISK FOR ZINC DEFICIENCY

Zinc deficiency in humans has been first described in the 1960s by Prasad and colleagues [138]. Since then, it was found to be a major health issue throughout the world. According to the World Health Organization (WHO), zinc deficiency in developing countries with high mortality is the fifth leading cause for the loss of healthy live years [189].

Diagnosis and Symptoms

The determination of zinc deficiency is a complicated and controversial issue. At present, a reliable and generally accepted method still needs to be found [114]. Generally, plasma or serum zinc concentrations are determined to assess zinc status, but plasma contains less than 0.1 % of total body zinc [158]. In addition to diurnal variations, it is affected by infections, fever, food intake, pregnancy, starvation, and catabolism [118, 158]. Hence, this method only allows the diagnosis of a pronounced lack of zinc.

Many biological processes are affected by zinc deficiency and its symptoms are unspecific. They include alopecia, dermatitis, decreased wound healing, neurological disturbances, and increased susceptibility to infections [118, 132, 138]. None of these symptoms is exclusive for zinc deficiency, and an unambiguous diagnosis based on the clinical presentation is difficult.

Subclinical Zinc Deficiency

Subclinical zinc deficiency is not limited to developing countries, it occurs also with relatively high incidence in industrialized parts of the world. Here, it can either result from insufficient nutrition, or as secondary zinc deficiency as a consequence of diseases which cause impaired intestinal absorption and/or an increase in intestinal loss of zinc [118]. Prominent examples include patients suffering from sprue, cystic fibrosis, chronic bowel disease, diabetes, renal insufficiency, cirrhosis of the liver, alcoholism, stress, and catabolism [87, 118].

Populations at risk are particularly children and adolescents, pregnant women, alcoholics, and the elderly [82]. In studies with elderly in the USA, more than half of the participants had a non-adequate zinc uptake [15, 142]. However, it has to be noted that these figures were based on Recommended

Daily Allowances (RDA) which are no longer in use (former U.S. RDA: 15 mg/d male and 12 mg/d female) [68]. Currently, the U.S. recommendations for zinc intake are 11 mg/d for males and 8 mg/d for females [1].

Manifest and Hereditary Zinc Deficiency

Malnutrition is a major cause for zinc deficiency. The Food and Agricultural Organization (FAO) calculated on the basis of the worldwide food supply that nearly half of the world's population is at risk for insufficient zinc intake. For Europe and North America the risk is between 1 and 13 %, for South and South-eastern Asia, Africa, and the Eastern Mediterranean the risk ranges from 68 to 95 % [17].

A genetic defect is responsible for the autosomal-recessive inherited disease *Acrodermatitis enteropathica*. Here, a mutation in the intestinal zinc transport protein Zip4 limits the uptake of zinc from food [123]. Symptoms start after weaning and include dermatitis, alopecia, thymic atrophy, and diarrhea. Without treatment, the children die within a few years, but the disease can be effectively treated by administration of high zinc doses [2].

Other genetic defects are known to affect newborns from mice and humans [2]. Due to mutations in the zinc transport proteins that supply zinc to the mother's milk (in humans ZnT2 [27]), this milk of the zinc-sufficient mothers does not contain enough zinc, leading to zinc deficiency in the offspring. This can have severe consequences; the mouse strain with a comparable mutation has become known as the ´lethal milk mouse`. The newborn are treated successfully by oral zinc supplementation until weaning [2].

Zinc Supplements and Bioavailability

Zinc can be found in foodstuffs from animal and plant sources. However, zinc content and bioavailability are generally higher in meat. Some food plants, such as corn, contain high amounts of chelators such as phytate, which inhibits zinc uptake by forming a zinc/phytate complex with extremely low bioavailability. The uptake of zinc from food can be complemented by zinc supplements. These vary considerably with regard to their elemental zinc content and its bioavailability (Table 1).

Table 1. Comparison of several zinc supplements

Bioavailability[a]	Supplement Name	Molecular formula[b]	Elemental zinc content [%][b,c]
high	zinc-L-aspartate	$Zn(C_4H_6NO_4)_2$	19.8
	zinc-L-cysteine	$Zn(C_3H_6NO_2S)_2$	21.4
	zinc-L-histidine	$Zn(C_6H_8N_3O_2)_2$	17.5
	zinc-L-methionine	$Zn(C_5H_{10}NO_2S)_2$	18.1
good	zinc chloride	$ZnCl_2$	48.0
	zinc gluconate	$Zn(C_6H_{11}O_7)_2$	14.4
	zinc sulphate	$ZnSO_4$	40.5
moderate	zinc acetate	$Zn(C_2H_3O_2)_2$	35.6
	zinc orotate	$Zn(C_5H_3N_2O_4)_2$	17.4
low	zinc oxide	ZnO	80.3

[a] Bioavailability cited from [16].
[b] Some zinc salts contain water molecules (e.g., $ZnSO_4$ x $7H_2O$), which leads to different molecular formulas and weight. Here, composition and zinc content are given for compounds without water.
[c] Elemental zinc content cited from [132].

Out of all existing supplements, most studies were performed with zinc sulphate. Bioavailability of the other the zinc salts is predominantly evaluated in comparison to zinc sulphate or, in some cases, zinc oxide. Moreover, enteric-coated preparations can be applied, as an increased zinc intake can cause gastric disorders [129, 157]. It is important to monitor the patient's copper status during supplementation with 30 mg zinc per day and above, as copper deficiency anemia has been observed due to zinc-induced changes in copper metabolism [136, 141, 156].

ZINC DEFICIENCY AND IMMUNE FUNCTION

The immune system protects the body from pathogens and the associated damage. It is composed of innate and adaptive immunity [175]. Macrophages, neutrophils and natural killer (NK) cells belong to innate immunity [86, 175]. Within minutes and hours after pathogens enter the body, they are recognized and destroyed by innate immune cells, and most pathogens are detected and destroyed by mechanisms relying exclusively on innate immunity, often without observable symptoms [175]. Neutrophils are short-lived cells, which are produced in great numbers and migrate to the origin of infection or inflammation [11]. They are the first line of defense, attacking pathogens by phagocytosis [11]. In a similar manner, macrophages eliminate invading pathogens, as well as dead cells and cell debris, by phagocytosis [62]. Furthermore, macrophages induce inflammation by the production of cytokines. The function of NK cells is to recognize and eliminate tumor and virus-infected cells [180].

T and B cells, which belong to adaptive immunity, become involved if an infection cannot be controlled by innate immunity alone [86]. Both cell types produce proteins suitable for the specific recognition of so-called antigens. The specificity is unique for each cell [83] and originates from somatic recombination of the coding DNA during maturation [30, 134]. B cells produce antibodies [171], which bind to the pathogens and toxins. Antibodies render the invaders recognizable for phagocytic cells and activate the complement system. The complement system consists of a large number of plasma proteins and is activated by binding to antigen-antibody complexes or pathogen surfaces [11]. The activation results in killing of the pathogens by forming a pore in their outer membrane, by recruitment of immune cells, and by opsonization (facilitation of phagocytosis) [11, 195]. In contrast to B cells, T cells are only activated following the assimilation, processing and presentation of antigens by antigen-presenting cells, such as dendritic cells and macrophages [172]. The T cell subtypes have different functions. Cytotoxic T cells eliminate virus-infected cells and tumor cells [185]. T helper cells (Th cells) facilitate and support the functions of other cells [64]. E.g., Th1 cells activate infected macrophages to kill phagocytozed pathogens, whereas Th2 cells support antibody production by B cells [64]. Another important T cell subtype are regulatory T cells, which diminish the activities of other immune cells, thereby limiting the immune response [165].

Adaptive immunity is induced with a delay of several days after the first contact with a pathogen [175], because it depends on transport of the antigens to lymphoid tissues, where they are recognized by naïve B and T cells that express receptors specific for these particular antigens. Thereby, the B and T cells are activated, proliferate and differentiate into mature effector cells [30]. This leads to the formation of a large number of antigen specific adaptive immune cells, which can counteract the present infection with high efficiency, but also form memory cells, which allow a faster and more powerful response upon subsequent contacts with the same pathogen [172]. This latter effect is mediated by so-called memory cells, and forms the basis for building a protective immunity by vaccination [19].

Innate and adaptive immunity do not act independently; instead the immune response is a concerted action of multiple components [11, 104, 175]. To coordinate this action, immune cells produce messenger proteins, so-called cytokines, which function in auto-, para- and sometimes endocrine fashion [50].

Zinc is indispensible for an effective immune response, being involved in multiple functions. For example, the secretion of cytokines is affected by an individual's zinc status. Additionally, the immune response depends on the efficient proliferation and differentiation of its various cellular components. Hence, the immune system, like other strongly proliferating organs such as skin and reproductive organs, depends on sufficient zinc availability [115, 150, 155]. Moreover, zinc is a regulator in signal transduction and apoptosis, and therefore involved in the control of the immune response on the cellular level [71, 135].

Zinc homeostasis is controlled by different mechanisms. There are two families of zinc transporter proteins. The Zrt- and Irt-like proteins (Zip) transport zinc from extracellular fluid or intracellular compartments into the cytosol, whereas the solute-linked carrier 30 (ZnT) proteins transport zinc in the opposite direction [33, 151]. Expression of zinc transporters is regulated by dietary factors, cytokines and hormones [111]. Within cells zinc can be stored in vesicles, so-called zincosomes [184]. Another mechanism for regulating zinc homeostasis is based on the cytoplasmic protein metallothionin (MT), which can bind up to seven zinc ions [103, 117]. Zinc can be mobilized from MT by thiol-oxidation of its zinc-binding cysteine side chains. Zinc release from its different storage sites leads to elevated concentrations of free zinc, which is involved in several signal transduction pathways of immune cells [70].

Zinc and Immune Cells

Neutrophils

In the case of an infection, neutrophils are the first cells to actively enter the infected tissue and come into contact with the pathogens [187]. Zinc deficiency as well as excess zinc affects the functions of neutrophils (Table 2). The general requirement of zinc for proliferation (see below) is particularly important for neutrophils, as 1-2 x 10^{11} cells per day of these short-lived cells are produced in the bone marrow [150, 187]. Neutrophils get to the infected tissue in a process called chemotaxis. Therefore, the cells follow a concentration gradient of messenger substances that are formed at the site of infection. During zinc deficiency, the chemotactic response of neutrophils is reduced [150, 163]. On the other hand, zinc concentrations of ≥ 500 μM can induce chemotactic activity of neutrophils *in vitro* [80]. At the origin of infection, neutrophils take up pathogens by phagocytosis and kill them by oxidative burst through the production of reactive oxygen species, such as hydrogen peroxide, superoxide anion and hypochlorous acid. Superoxide anion is the primary substance formed during the oxidative burst. It is produced by the zinc-sensitive enzyme NADPH oxidase. *In vitro*, zinc deficiency as well as zinc excess inhibits the NADPH oxidase of neutrophils [38, 74]. Conversely, the addition of moderate zinc concentrations can increase the enzyme's activity [60, 74]. Notably, this effect is also supported by *in vivo* observations. Neutrophils from elderly persons have a decreased zinc content and there is a clear reduction in the oxidative burst [128]. Phagocytosis of neutrophils and macrophages was also shown to be decreased during zinc deficiency, and increased by zinc supplementation [93, 164]. One study with peripheral blood mononuclear cells (PBMC) from canines proposed that the effect is mediated by the altered secretion of the cytokine tumor necrosis factor-α (TNF-α) by monocytes, which acts in an auto- and paracrine fashion on phagocytosis [94].

Table 2. Impact of zinc levels on neutrophil functions

Zinc deficiency
• Chemotactic response ↓ [150, 188] • Phagocytosis ↓ [93] • Oxidative burst ↓ [74] → Susceptibility to infection ↑ [16]
Moderate zinc supplementation
• Phagocytic activity restored [93, 164] • NADPH oxidase activity ↑ [60, 74]   ○ superoxide anion formation ↑ [60, 74] → Successful termination of infection
Excess zinc
• ≥ 100 μM: NADPH oxidase activity ↓ [38]   ○ superoxide anion formation ↓ [38] → killing of pathogens ↓ • > 500 μM: chemotactic response induced [80] → Positive effects in skin cream [16]

Monocytes/Macrophages

Monocytes are found in the blood, and extravasate into the tissue, where they mature into resident macrophages. During infection, additional monocytes are recruited to the affected tissue through chemotaxis. This involves adhering to endothelial cells as a first step of entering the infected tissue. *In vitro*, adhesion of monocytic cells is increased by supplementation of physiological zinc concentrations and decreased by chelation of zinc [24]. Stimulation of monocytes with monocyte chemoattractant protein-1 (MCP-1) or phorbol myristate acetate (PMA) induces an intracellular increase of free zinc, which augments the adhesion to endothelial cells [100]. Besides the direct effects on monocytes, expression of intercellular adhesion molecule-1 (ICAM-1), an adhesion molecule of the endothelium, is also increased by *in vitro* supplementation with zinc [173, 193].

Table 3. Impact of zinc levels on monocyte/macrophage functions

Zinc deficiency
• Vitamin D3-dependent differentiation ↑ [44]
• Chemotactic response ↓ [163]
• Adhesion ↓ [24]
• Phagocytosis ↓ [93, 164]
→ Increased susceptibility to infection [16]
Moderate zinc supplementation
• Vitamin D3-dependent differentiation ↓ [44]
• Adhesion ↑ [24]
• Phagocytosis ↑ [93, 164]
• Supplementation with
o 1 μM $ZnSO_4$ (+ ionophore pyrithione): TNF-α ↑ [181]
o 25 μM $ZnSO_4$ (+ ionophore pyrithione): TNF-α, IL-1β ↓ [181]
• LPS-induced IL-6, IL-8 and IL-10 release ↑ [90]
• Basal IL-1β, IL-6 and IL-8 release ↓ [90]
→ Improvement of chronic inflammation
→ Successful termination of infection
High zinc doses
• ≥ 100 μM: direct activation of monocytes and macrophages [150]
• ≥ 500 μM: direct activation and TNF-α and IL-1β secretion [42]
→ Positive effects in skin cream [16]

After macrophages have been activated, they take up pathogens or cellular debris by phagocytosis and kill the pathogens by oxidative burst. Both functions are influenced by zinc levels as discussed earlier in this chapter. One way by which monocytes and macrophages recognize pathogens is through so-called Toll-like receptors (TLRs). Stimulation of TLR-4 with the endotoxin lipopolysaccharide (LPS) generates a zinc signal in the cytoplasm of monocytes within 2 minutes [69]. This signal is required for activation of several TLR-4 dependent signaling pathways, including the transcription factor NFκB and mitogen activated protein kinases, which are involved in the release of pro-inflammatory cytokines [69]. Table 3 summarizes the impact of zinc on monocyte and macrophage functions.

NK Cells

NK cells are lymphocytes that belong to innate immunity. They act by two main mechanisms. Firstly, NK cells sense alterations in the expression of major histocompatibility complex class-I (MHC-I) molecules and some other cell-surface proteins. In case of missing or reduced MHC-I presentation, NK cells release the content of cytotoxic granules, killing their target cells. This is necessary because reducing the expression of MHC-I molecules on their surface is a mechanism for evading immune surveillance. Normally, cells present antigens from their cytosol on MHC-I proteins, which is a means of controlling for viral or tumor antigens by cytotoxic T cells. The second mechanism is called antibody-dependent cellular cytotoxicity (ADCC). NK cells carry immunoglobulin receptors that bind to the constant regions of antibodies. The presence of antibodies on the cell surface results from the detection of antigens indicative of viral infection or malignant transformation, activating the NK cells to release the content of cytotoxic granules.

NK cells also depend on zinc (Table 4). Here, extracellular zinc is required for MHC-I recognition on potential target cells [147, 148]. Hence, unspecific killing might increase during zinc deficiency [150]. Conversely, the lytic activity of NK cells is decreased during deficiency. It has been suggested that decreased secretion of the cytokine interleukin-2 (IL-2) by T cells is one mechanism responsible for reduced lytic activity of NK cells [137]. *In vitro*, physiological zinc levels (10 μM) increase the lytic activity, proliferation, and differentiation of purified NK cells from young and old donors. Higher zinc concentrations (20 μM) reduce the proliferation of NK cells from elderly, underlining the importance of physiological zinc levels on NK cell development [130].

Table 4. Impact of zinc levels on NK cell functions

Zinc deficiency
• Inhibitory signals ↓ [147, 148]   o unspecific killing ↑ [150, 152] • Lytic activity ↓ [137] → Increased susceptibility to infection [16]
Moderate zinc supplementation
• Lytic activity ↑ [130] • Proliferation and differentiation ↑ [130] → Successful elimination of infection
Excess zinc
• 50-100 μM: cytotoxicity ↓ [54] → Tumor development possibly facilitated [16]   o treatment with zinc doses > 100 mg/day potentially increases prostate cancer risk, mechanism is unknown [88, 107]

T Cells

T cells are particularly vulnerable to zinc deficiency [96], as illustrated in Table 5. The deficit affects both T cell development and function [77]. T cell progenitors migrate from the bone marrow into the thymus where they undergo a process of maturation. Chronic and acute zinc deficiency causes thymic atrophy and T cell lymphopenia [57]. In the maturation process, T cells are especially vulnerable during their phase as pre-T cells. A loss of 50 % of the cells in this stage resulted from zinc deficiency in mice [96]. T cells in other maturation phases (pro-T cells, T helper cells and cytotoxic T cells) and other leukocytes (B cells, monocytes, and neutrophiles) are more resistant to zinc deficiency-induced apoptosis [96]. Elevated glucocorticoid levels caused by zinc deficiency and different expression of the antiapoptotic protein Bcl-2 in the maturation stages are probably involved in the increased apoptosis of pre-T cells [57]. The peptide hormone thymulin is an additional factor during T cell maturation. Zinc is a co-factor for thymulin, which is biologically inactive in the absence of zinc [72, 143]. During zinc deficiency, the serum thymulin activity is decreased, affecting T cell differentiation and function [137, 143]. This effect is reversed by zinc supplementation [137, 143].

A deficit of zinc affects T cell-dependent immune functions also through its impact on cell proliferation (see below). Moreover, zinc affects the balance of the different T cell subsets. A deficiency decreases the ratio of Th1 to Th2 cells, detected by the reduced secretion of the cytokines IFN-γ and IL-2 [10, 137]. Zinc supplementation restores the Th1 responses [10, 137]. The Th2 response is less affected by zinc deficiency [137], but it increases when zinc is supplemented [122]. A diet causing mild zinc deficiency reversibly decreases the ratio of T helper cells to cytotoxic T cells [10]. Additionally, mild zinc deficiency is associated with a decreased ratio of naïve to memory T lymphocytes, which is reversed during zinc repletion [10].

Table 5. Impact of zinc levels on T cell functions

Zinc deficiency

- Thymulin activity ↓ [143]
    - T cell differentiation and function ↓ [143]
- Thymus atrophy, apoptosis of pre-T cells ↑ [57, 96]
    - Lymphopenia [57, 96]
- Proliferation ↓ [91]
- T cell balance disturbed
    - Th1 cells ↓ [10, 137]
        - IFN-γ ↓, IL-2 ↓ [10, 137]
        - delayed type hypersensitivity reaction (Th1 dependent) in mice ↓ [10, 58]
    - Th2 ↔ [10, 137]
        - IL-4, IL-6, IL-10 ↔ [10, 137]
    - Ratio T helper cells to cytotoxic T cells ↓ [10]
    - Ratio naïve to memory T cells ↓ [10]
- Autoreactivity and alloreactivity ↑ [150]

→ Increased autoimmunity and susceptibility to infection [16]

Moderate zinc supplementation

- Proliferation ↑ [169]
- T cell balance
    - Reconstitution of Th1 / Th2 balance [10]
    - Reconstitution of T helper cells to cytotoxic T cells ratio [10]
    - Ratio of naïve to memory T cells ↑ [164]
- Th1 response ↑ in cholera vaccinated children, Th2 response ↔ [4]
- Superantigen-induced IFN-γ, TNF-α, IL-10 secretion ↑ [90]

→ Efficient activation of cellular immunity
→ Successful termination of infection

Excess zinc

- T lymphocyte function ↓ [22]
- Alloreactivity ↓ [18, 51]
- Development of regulatory Th17 cells ↓ [98]
    - Suppressive effect on autoimmunity? [98]

→ Susceptibility to infection increased [16]
→ Alloreactivity after transplantation suppressed [16]

B Cells

The impact of zinc on B cell development and function is less pronounced, compared to T cells (Table 6). During the maturation process in the bone marrow, B cells depend on zinc. In particular, pre- and immature B cells are vulnerable to a deficit [57]. Furthermore, zinc deficiency decreases antibody production [39, 56]. Mainly T cell-depending antibodies are affected [95, 163], implying that this is an effect on T helper cells rather than B cells [56]. The effects of zinc on B cells are particularly relevant in the context of vaccination. A correlation between decreased serum zinc levels and a declined success of vaccination has been noted in a study with hemodialysis patients [102]. The principle of vaccinations is the formation of antibodies and memory B cells through contact with the vaccine, which will accelerate a successful immune response in case of a second contact with the same antigens. In a large number of studies, the impact of zinc supplementation on vaccination of zinc-deficient individuals has been evaluated, with varying outcomes [4, 5, 45, 92, 145, 146, 174]. In some instances, extremely high zinc doses were administered, in one case up to 400 mg zinc sulphate per day [145]. These amounts suppress T helper cell activity and are thereby highly likely to impair the success of vaccination [152]. Most importantly, the time of zinc administration seems to be important. A start at the time of vaccination is ineffective [174], whereas supplementation that started in advance of vaccination was shown to have a better impact on the generation of protective antibodies [4, 5, 45]. This may be due to a positive impact on T helper cell development and function prior to vaccination, which will then result in improved T cell help to B cells during vaccination.

Table 6. Impact of zinc deficiency on B cell functions

Zinc deficiency
• B lymphocyte development ↓ [57]
○ depletion of the naïve B lymphocyte fraction [57]
• Antibody production ↓ [39]
○ IgM and IgD ↔, IgG and IgA ↓ [95, 163]
→ Potentially reduced response to vaccination [39]

Cell Proliferation

Efficient proliferation is essential for a successful immune response. Zinc affects proliferation on different levels. Firstly, zinc is required for DNA and protein synthesis [115], as proteins involved in DNA and RNA synthesis, e.g.

RNA polymerases, are zinc metalloproteins [29, 115, 192]. Furthermore, the activity of thymidine kinase, which is important for the incorporation of thymidine into DNA and therefore DNA synthesis, is reduced during zinc deficiency [25, 115, 139]. In a T helper cell line, the reduced activity of thymidine kinase in the G1 and S phase of the cell cycle has been shown to cause a delay in the cell cycle [139]. This was based, among other effects, on a decreased transcription of thymidine kinase mRNA [115, 139].

Additionally, zinc induces alterations on the endocrine level, affecting cell proliferation. Growth factors involved in proliferation, such as growth hormone (GH) and Insulin-like growth factor-I (IGF-I), are known to be sensitive to the zinc status [115]. Feeding rats with a zinc-deficit diet for several days results in decreased circulating GH and IGF-I concentrations [153]. Accordingly, decreased body weight gain has been observed in comparison to pair-fed rats receiving the same amount of a zinc-sufficient diet [153].

Finally, intracellular free zinc signals participate in regulating cell proliferation. In this context, the cytokine IL-2 is interesting. One general function of the cytokine is to induce proliferation of T cells [61]. *In vitro* experiments have shown that the IL-2-induced activation of mitogen-induced protein kinases and the resulting proliferation were reduced after blocking IL-2-induced zinc signals with a chelator [91].

Inflammation and Cytokines

Inflammation is an essential immune function [126]. It is a complex reaction that enables survival during infection and injury [126]. Triggers are infections, tissue injuries, and other types of adverse conditions, e.g. obesity [126]. Tissue-resident macrophages and mast cells respond with the production of inflammatory mediators such as cytokines (Table 7), chemokines, and eicosanoids. These induce an influx of plasma proteins and the migration of leukocytes into the inflamed tissue. A successful termination is important, as the inflammatory response can shift from acute to chronic [126]. Chronic inflammation occurs in a variety of diseases, for example in diabetes type 2, atherosclerosis, asthma and cancer [125, 126]. Taken together, controlled inflammation is beneficial, as it protects against infections, but an imbalanced response is harmful [125].

The trace element zinc is critical for inflammation, because it affects the release of several pro-inflammatory cytokines. For example, the supplementation of elderly with 10 mg zinc daily for nearly seven weeks reduced the basal pro-inflammatory cytokine secretion of isolated peripheral blood mononuclear cells (PBMC), indicating positive effects on basal inflammation [90]. Monocytes and macrophages produce pro-inflammatory cytokines when LPS binds to the TLR-4. Simultaneous stimulation with LPS and zinc results in an altered release of pro-inflammatory cytokines, and the resulting effects depend on the concentration of zinc [181]. Moderate amounts augment cytokine production [41, 43], whereas higher zinc concentrations decrease the release of TNF-α and IL-1β [181]. Additionally, cytokine release of T helper cells is also affected by the zinc status. The Th1 cytokines interferon- γ (IFN-γ) and TNF-α are reduced, whereas the Th2 cytokines IL-4 and IL-10 remain unaltered as depicted in Figure 1 [10, 137].

Table 7. Zinc-modulated pro- and anti-inflammatory cytokines

Cytokine	Producing cells	Function
Pro-inflammatory cytokines		
IFN-γ	Th1 cells cytotoxic T cells NK cells	•promotes differentiation of Th0 cells to Th1 cells •inhibits Th2 cell proliferation •activates macrophages and stimulates synthesis of inflammatory mediators •induces immunoglobulin class switch in B cells •activates cytotoxic T cells and NK cells
TNF-α	macrophages NK cells Th1 cells	•activates macrophages and stimulates synthesis of inflammatory mediators •activates cytotoxic T cells, NK cells and endothelial cells •fever
IL-1β	macrophages epithelial cells	•activates cytotoxic T cells, macrophages, NK cells and endothelial cells •induces IL-6 production •fever
IL-2	Th1 cells	•T cell and NK cell proliferation
IL-6	T cells macrophages endothelial cells	•T- and B-cell proliferation and activation •acute phase proteins •fever •increases antibody production
IL-8 (CXCL8)	macrophages endothelial cells	•mobilizes, activates and causes degranulation of neutrophiles
Anti-inflammatory cytokines		
IL-4	Th2 cells	•activates B cells and induces proliferation •induces immunoglobulin class switch •induces differentiation of Th0 to Th2 cells •inhibits macrophage activation
IL-10	Th2 cells monocytes	•inhibits Th1 cells •inhibits macrophage functions, e.g. cytokine release

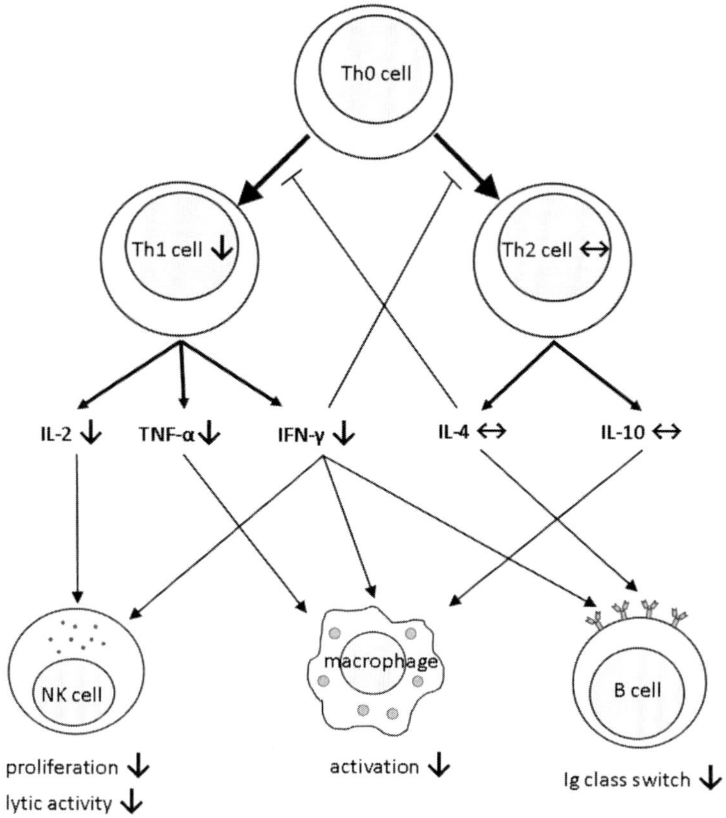

Figure 1. Zinc deficiency shifts the T helper cell response. The differentiation of Th0 into Th1 cells is decreased by a zinc deficit, whereas differentiation of Th0 into Th2 cells remains unaltered. Hence, the Th1 cytokines IL-2, TNF-α, and IFN-γ are reduced and the Th2 cytokines IL-4 and IL-10 remain unchanged. The lack of IL-2, TNF-α, and IFN-γ causes decreased proliferation and lytic activity of NK cells, decreased activity of macrophages and reduced Ig class switch of B cells (↓ reduced by zinc deficiency; ↔ unaffected by zinc deficiency).

During the early stages of an infection, cytokine release induces alterations in plasma protein concentrations which are known as acute phase response. Simultaneously, the plasma zinc concentration decreases (Figure 2). This is caused by the pro-inflammatory cytokines IL-6 and TNF-α secreted by macrophages and monocytes. TNF-α upregulates Zip8 on monocytes and lung epithelia cells, initiating a zinc influx [99]. IL-6 upregulates Zip14 on hepatocytes, triggering zinc uptake and intracellular storage bound to metallothionin (MT) [113]. Most likely, the shift of zinc into the liver serves the purpose of depriving pathogens of this essential trace element, thereby inhibiting their growth [32].

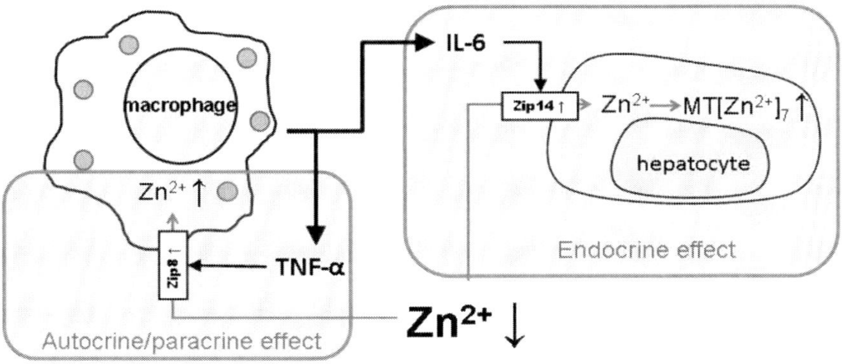

Figure 2. Plasma zinc concentrations decrease during the acute phase. Macrophages release TNF-α and IL-6 during an inflammatory immune response. In an auto- and paracrine fashion, TNF-α induces the expression of the zinc transporter Zip8, resulting in increased zinc uptake by macrophages. In addition, hepatocytes upregulate Zip14 expression in response to IL-6, followed by increased zinc uptake and intracellular storage bound to metallothionin (MT). Overall, the release of pro-inflammatory cytokines causes a decline of plasma zinc levels, mainly resulting from a translocation into the liver (↓ reduced during the acute phase; ↑ elevated during the acute phase).

Sepsis is a systemic inflammatory response to infection that includes sepsis-associated organ-dysfunction [108]. Acute circulatory failure, which is characterized by persistent arterial hypotension, emerges during septic shock [108]. Septic shock is associated with altered zinc homeostasis, as lower serum zinc levels in septic children are predominantly associated with non-survivors of septic shock compared to survivors [191]. Septic patients show lower plasma zinc contents and higher plasma cytokine concentrations compared to non-infected critically ill subjects. Furthermore, increased severity of illness correlates to low plasma zinc and enhanced expression of zinc transporter Zip8 in monocytes [99].

Zinc Status and Autoimmunity

Zinc deficiency is associated with several autoimmune diseases such as inflammatory bowel disease [182], rheumatoid arthritis [132, 166], and diabetes type I (see below). The underlying mechanisms, however, are still unclear. Presumably, the association between zinc and autoimmune disease is based on the interplay of zinc and T cell activity (Table 5). A subset of T helper cells, the so-called Th17 cells, is a major contributor to autoimmune inflammation [7, 101]. The cytokine IL-17 that is produced by Th17 cells induces the secretion of chemokines and migration of neutrophils into infected tissue [101]. Zinc applied in high concentrations in the drinking water suppresses the development of arthritis in a mouse model, by interfering with the signaling pathway of IL-6 [98]. Together with transforming growth factor-β, IL-6 induces the differentiation of naïve T cells into Th17 cells [7].

DISEASES AND ZINC SUPPLEMENTATION

Diarrhea

With 1.6 million deaths in 2002, diarrhea is a leading cause of child mortality [190]. During the course of this infectious disease, poor intestinal absorption leads to a lack of essential nutrients, including zinc [183]. In malnourished children, zinc malabsorption caused by diarrhea results in progression of zinc deficiency and impairs cellular immunity. Consequently, episodes and duration of diarrhea increase and the children enter a vicious circle of zinc deficiency and infection [183]. Zinc supplementation is an effective tool to escape this circle. Long-term zinc supplementation for at least three months reduced the frequency, severity, and duration of diarrhea in children younger than five years [3]. Additionally, short-term zinc supplementation was also effective [12]. 20 mg zinc daily for 2 weeks reduced diarrhea incidence and prevalence in children younger than 5 years old in developing countries, similar to continuous supplementation [12]. Zinc treatment after onset of diarrhea is also comprehensively studied [55]. In children older than six months zinc supplementation reduced the duration of acute and persistent diarrhea [106]. Accordingly, the WHO recommends the supplementation of sick children younger than five years with 20 mg zinc daily for 10-14 days (10 mg daily for children younger than six months) [190].

Pneumonia

Among the infectious diseases, pneumonia is the leading cause of death in children younger than five years [13]. Hence, the beneficial impact of zinc supplementation on the progress of infection is of great importance.

Prophylactic zinc supplementation of children younger than five years is beneficial [105]. 10 mg zinc daily for 4-20 months reduced the incidence and prevalence of pneumonia [105]. The success of zinc administration after disease outbreak is less clear [73]. Zinc supplementation in addition to the standard treatment reduced the duration of severe pneumonia only in one study [73], in contrast to six other trials that did not detect this effect [8, 73, 176].

So far, the investigation of zinc supplementation on the progress of pneumonia concentrated on malnourished children. The high frequency of an insufficient zinc status in elderly might also be a risk factor for the development of pneumonia, considering the high prevalence of the disease in this age group [9].

Common Cold

There is a long and controversial discussion whether zinc treatment decreases the severity and duration of the common cold [81, 84, 120, 121]. A recent Cochrane meta-analysis came to the conclusion that duration and severity of the common cold in otherwise healthy people is reduced by zinc, if it is administered in the form of syrup or lozenges within 24 hours of the onset of symptoms [167]. In this analysis, the daily doses ranged from 30 mg (syrup) to 90-160 mg (lozenges) and were given for 5-14 days or as long as the symptoms persisted [167]. Zinc treatment, in particular zinc lozenges, caused significantly more adverse side effects than placebo treatment, such as bad taste and nausea [36]. It has been suggested that zinc ions are the beneficial compound [47], and the varying release of ionic zinc from different zinc salts is the likely reason for the differences and uncertainties observed in the different trials [47].

Diabetes

Diabetes mellitus is a disease of worldwide prevalence. Two main forms can be distinguished. Type 1 diabetes is characterized by a lack of insulin production. Here, the insulin-secreting beta-cells of the pancreas are destroyed through autoimmune mechanisms and patients depend on insulin treatment for the rest of their lives [14]. The more widespread form is type 2 diabetes. In 2010, the estimated worldwide incidence in adults was as high as 285 million [131]. Type 2 diabetes is characterized by hyperglycemia and alterations in the lipid metabolism. Insulin resistance of tissues causes increased insulin production of beta-cells. Eventually, the function of beta-cells declines, resulting in insufficient insulin production and hyperglycemia.

Diabetes mellitus is associated with zinc metabolism. Firstly, diabetic patients have an increased excretion of zinc in the urine, causing hyperzincuria [20, 97]. In accordance, zinc concentrations in leucocytes decrease in both types of diabetes [87]. In contrast, the variation of serum and plasma levels seems to depend on the diabetes type, being increased in type 1 but decreased in type 2 [87]. Secondly, zinc is needed for storage of insulin in beta-cells [23, 162]. The zinc transport protein ZnT8 mainly transports zinc into the secretory vesicles [26]. Notably, autoantibodies against ZnT8 have recently been identified as a useful marker for type 1 diabetes [178, 186] and there is an association between a polymorphism in the gene coding for ZnT8 and the risk for developing type 2 diabetes [21, 168]. In addition, a polymorphism in the gene encoding the intracellular zinc storage protein MT2A might also be associated with type 2 diabetes. Patients carrying the AA and not the AG genotype at locus -209 within the MT2A promoter region show higher IL-6 levels, hyperglycaemia, enhanced glycated haemoglobin (HbA1c) levels and increased zinc deficiency [63]. Among the pro-inflammatory cytokines, particularly IL-1β is associated with diabetes. It affects the conversion of proinsulin into the mature form [78] and glucose-stimulated insulin secretion [116]. It is also involved in the islet cell destruction in type 1 and type 2 diabetes [28, 116]. The expression of zinc transporters, in particular of ZnT8, is sensitive to IL-1β and other pro-inflammatory cytokines [48, 49]. Hence, zinc interferes with the development of diabetes, at least to some extent, through IL-1β [87]. Additionally, zinc exhibits insulin-like activity and is involved in insulin-triggered intracellular signal transduction by inhibiting protein tyrosine phosphatases [67, 124,170].

The impact of zinc supplementation on diabetic patients has been examined repeatedly [34, 37, 52, 149, 194]. In patients with type 1 diabetes the outcomes are controversial [34, 37, 52] and it has to be noted that HbA1c increased in two studies, indicating an undesired enhancement of plasma glucose over a prolonged period of time [34, 37]. Supplementation of type 2 diabetes patients is possibly beneficial as determined by decreased oxidative stress [154] and decreased fasting plasma glucose, insulin levels and insulin resistance [75].

Cancer

In recent years, an involvement of zinc in the development and progression of cancer has become obvious [59]. Zinc exhibits systemic effects, for example on cytokine release, which affects tumor cell growth [59, 66]. On the cellular level, zinc is an important component of DNA binding proteins and proteins involved in DNA repair [76]. Additionally, zinc affects cell proliferation and several intracellular signaling pathways important for apoptosis. The transcription factor p53 is important for an adequate cellular response to DNA damage. It detects the damage and initiates apoptosis when necessary [127]. Zinc influences the conformation of this transcription factor by stabilizing the tertiary structure of its DNA binding domain [179] and induces p53 protein expression [144]. Zinc also affects apoptosis through cytochrome c release and mitochondrial apoptogenesis [53].

Altered zinc concentrations in cancer tissue were repeatedly observed. Tissue zinc concentrations are markedly increased in breast cancer [119, 159]. Aggressiveness and metastatic activity of the malignant cells are associated with the expression of the zinc transporter ZIP10 [89]. Notably, preventive zinc supplementation for longer than ten years has protected Canadian pre- and postmenopausal women from the development of breast cancer [133]. Zinc uptake and accumulation is also increased in pancreas cancer cells through enhanced expression of Zip4 [109]. High zinc concentrations supposedly induce growth promoting effects in malignant cells [40, 59, 109]. On the other hand, prostate cells, which normally contain high levels of intracellular zinc, lose their ability to accumulate zinc as a result of malignant transformation [31, 59, 79]. Exposure of prostate cancer cells to zinc inhibits tumor growth to some extent through induction of apoptosis [110]. The intake of more than 15 mg supplemental zinc daily for ten years has been associated with a decreased risk for advanced prostate cancer in the VITamins And Lifestyle (VITAL) cohort, which includes 35,242 men [65]. A daily zinc intake of greater 100 mg is, however, suspected to increase the risk for developing prostate cancer [107]. In head and neck cancer patients zinc deficiency, measured as zinc content in lymphocytes and granulocytes, correlated to increased tumor size, higher tumor stage, less months of disease free intervals, higher percentage of morbidities, and higher numbers of unplanned stays in the hospital [140]. Other cancers, for example, cervical cancer [35], lung cancer [160], ovarian cancer [112], and hepatocellular cancer [46] are also associated with altered zinc metabolism.

Conclusion

The knowledge about the impact of zinc supplementation on immunity is steadily increasing. Still, there are many contradictory results, for example the impact of zinc on vaccination. Here, differences in study design, such as varying duration or administration of different zinc salts, as well as the lack of a sensitive marker for detecting marginal zinc deficiency have to be considered. Hardly any two studies were performed under entirely comparable circumstances.

It is, however, undisputed that zinc is an important nutrient for the rapidly proliferating cells of the immune system. The trace element affects differentiation, function, and activity of these cells. Thereby, zinc is involved in the formation of inflammation and autoimmunity and the development and progression of diseases. Several diseases are associated with an inadequate zinc status. Additionally, sub-clinical zinc deficiency, which occurs frequently even in industrialized countries, can result in impaired immune function.

Based on the vast impact of zinc on immunity discussed in this chapter, we consider a supplementation with up to 10 mg daily reasonable for patients with supposed zinc deficiency. At this dose no adverse effects are to be expected, whereas patients belonging to risk populations, such as the elderly, might benefit greatly from the impact on their immune system.

References

[1] A Report of the Panel on Micronutrients SoURLoNaoIaUoDRI, and the Standing Committee on the Scientific Evaluation of Dietary Reference Intakes. DIETARY REFERENCE INTAKES FOR Vitamin A, Vitamin K, Arsenic, Boron, Chromium, Copper, Iodine, Iron, Manganese, Molybdenum, Nickel, Silicon, Vanadium, and Zinc. In: *Food and Nutrition Board IoM,* editor. Washington, D.C.: NATIONAL ACADEMY PRESS; 2001.

[2] Ackland ML, Michalczyk A. Zinc deficiency and its inherited disorders - a review. *Genes Nutr.* 2006, 1, 41-49.

[3] Aggarwal R, Sentz J, Miller MA. Role of zinc administration in prevention of childhood diarrhea and respiratory illnesses: a meta-analysis. *Pediatrics.* 2007, 119, 1120-1130.

[4] Ahmed T, Arifuzzaman M, Lebens M, Qadri F, Lundgren A. CD4+ T-cell responses to an oral inactivated cholera vaccine in young children in a cholera endemic country and the enhancing effect of zinc supplementation. *Vaccine.* 2009, 28, 422-429.

[5] Ahmed T, Svennerholm AM, Al Tarique A, Sultana GN, Qadri F. Enhanced immunogenicity of an oral inactivated cholera vaccine in infants in Bangladesh obtained by zinc supplementation and by temporary withholding breast-feeding. *Vaccine.* 2009, 27, 1433-1439.

[6] Andreini C, Banci L, Bertini I, Rosato A. Counting the zinc-proteins encoded in the human genome. *J. Proteome. Res.* 2006, 5, 196-201.

[7] Awasthi A, Kuchroo VK. Th17 cells: from precursors to players in inflammation and infection. *Int. Immunol.* 2009, 21, 489-498.

[8] Bansal A, Parmar VR, Basu S, Kaur J, Jain S, Saha A, Chawla D. Zinc supplementation in severe acute lower respiratory tract infection in children: a triple-blind randomized placebo controlled trial. *Indian J. Pediatr.* 2011, 78, 33-37.

[9] Barnett JB, Hamer DH, Meydani SN. Low zinc status: a new risk factor for pneumonia in the elderly? *Nutr. Rev.* 2010, 68, 30-37.

[10] Beck FW, Prasad AS, Kaplan J, Fitzgerald JT, Brewer GJ. Changes in cytokine production and T cell subpopulations in experimentally induced zinc-deficient humans. *Am. J. Physiol.* 1997, 272, E1002-1007.

[11] Beutler B. Innate immunity: an overview. *Mol. Immunol.* 2004, 40, 845-859.

[12] Bhutta ZA, Black RE, Brown KH, Gardner JM, Gore S, Hidayat A, Khatun F, Martorell R, Ninh NX, Penny ME, Rosado JL, Roy SK, Ruel M, Sazawal S, Shankar A. Prevention of diarrhea and pneumonia by zinc supplementation in children in developing countries: pooled analysis of randomized controlled trials. Zinc Investigators' Collaborative Group. *J. Pediatr.* 1999, 135, 689-697.

[13] Black RE, Cousens S, Johnson HL, Lawn JE, Rudan I, Bassani DG, Jha P, Campbell H, Walker CF, Cibulskis R, Eisele T, Liu L, Mathers C. Global, regional, and national causes of child mortality in 2008: a systematic analysis. *Lancet.* 2010, 375, 1969-1987.

[14] Bluestone JA, Herold K, Eisenbarth G. Genetics, pathogenesis and clinical interventions in type 1 diabetes. *Nature.* 2010, 464, 1293-1300.

[15] Briefel RR, Bialostosky K, Kennedy-Stephenson J, McDowell MA, Ervin RB, Wright JD. Zinc intake of the U.S. population: findings from the third National Health and Nutrition Examination Survey, 1988-1994. *J. Nutr.* 2000, 130, 1367S-1373S.

[16] Brieger A, Rink L. Zink und Immunfunktionen. *Ernährung & Medizin.* 2010, 25, 156-160.

[17] Brown KH, Wuehler, S.E., Peerson, J.M. The importance of zinc in human nutrition and estimation of the global prevalence of zinc deficiency. *Food Nutr. Bull.* 2001, 22, 113-125.

[18] Campo CA, Wellinghausen N, Faber C, Fischer A, Rink L. Zinc inhibits the mixed lymphocyte culture. *Biol. Trace Elem. Res.* 2001, 79, 15-22.

[19] Campos M, Godson DL. The effectiveness and limitations of immune memory: understanding protective immune responses. *Int. J. Parasitol.* 2003, 33, 655-661.

[20] Canfield WK, Hambidge KM, Johnson LK. Zinc nutriture in type I diabetes mellitus: relationship to growth measures and metabolic control. *J. Pediatr. Gastroenterol. Nutr.* 1984, 3, 577-584.

[21] Cauchi S, Del Guerra S, Choquet H, D'Aleo V, Groves CJ, Lupi R, McCarthy MI, Froguel P, Marchetti P. Meta-analysis and functional effects of the SLC30A8 rs13266634 polymorphism on isolated human pancreatic islets. *Mol. Genet. Metab.* 2010, 100, 77-82.

[22] Chandra RK. Excessive intake of zinc impairs immune responses. *JAMA.* 1984, 252, 1443-1446.

[23] Chausmer AB. Zinc, insulin and diabetes. *J. Am. Coll. Nutr.* 1998, 17, 109-115.

[24] Chavakis T, May AE, Preissner KT, Kanse SM. Molecular mechanisms of zinc-dependent leukocyte adhesion involving the urokinase receptor and beta2-integrins. *Blood.* 1999, 93, 2976-2983.

[25] Chesters JK, Petrie L, Travis AJ. A requirement for Zn2+ for the induction of thymidine kinase but not ornithine decarboxylase in 3T3 cells stimulated from quiescence. *Biochem. J.* 1990, 272, 525-527.

[26] Chimienti F, Devergnas S, Favier A, Seve M. Identification and cloning of a beta-cell-specific zinc transporter, ZnT-8, localized into insulin secretory granules. *Diabetes.* 2004, 53, 2330-2337.

[27] Chowanadisai W, Lonnerdal B, Kelleher SL. Identification of a mutation in SLC30A2 (ZnT-2) in women with low milk zinc concentration that results in transient neonatal zinc deficiency. *J. Biol. Chem.* 2006, 281, 39699-39707.

[28] Cnop M, Welsh N, Jonas JC, Jorns A, Lenzen S, Eizirik DL. Mechanisms of pancreatic beta-cell death in type 1 and type 2 diabetes: many differences, few similarities. *Diabetes.* 2005, 54 Suppl 2, S97-107.

[29] Coleman JE. Zinc proteins: enzymes, storage proteins, transcription factors, and replication proteins. *Annu. Rev. Biochem.* 1992, 61, 897-946.

[30] Cooper MD, Alder MN. The evolution of adaptive immune systems. *Cell.* 2006, 124, 815-822.

[31] Costello LC, Franklin RB, Feng P, Tan M, Bagasra O. Zinc and prostate cancer: a critical scientific, medical, and public interest issue (United States). *Cancer Causes Control.* 2005, 16, 901-915.

[32] Cousins RJ, Lichten LA. Zinc transporters. In: Rink L, editor. *Zinc in Human Health.* Amsterdam: IOS press; 2011. ISBN: 978-1-60750-815-1.

[33] Cousins RJ, Liuzzi JP, Lichten LA. Mammalian zinc transport, trafficking, and signals. *J. Biol. Chem.* 2006, 281, 24085-24089.

[34] Cunningham JJ, Fu A, Mearkle PL, Brown RG. Hyperzincuria in individuals with insulin-dependent diabetes mellitus: concurrent zinc status and the effect of high-dose zinc supplementation. *Metabolism.* 1994, 43, 1558-1562.

[35] Cunzhi H, Jiexian J, Xianwen Z, Jingang G, Shumin Z, Lili D. Serum and tissue levels of six trace elements and copper/zinc ratio in patients with cervical cancer and uterine myoma. *Biol. Trace Elem. Res.* 2003, 94, 113-122.

[36] Das RR, Singh M. Zinc lozenges for the common cold: Should we ignore the side-effects? *Med. Hypotheses.* 2011, 77, 308-309.

[37] de Sena KC, Arrais RF, das Gracas Almeida M, de Araujo DM, dos Santos MM, de Lima VT, de Fatima Campos Pedrosa L. Effects of zinc supplementation in patients with type 1 diabetes. *Biol. Trace Elem. Res.* 2005, 105, 1-9.

[38] DeCoursey TE, Morgan D, Cherny VV. The voltage dependence of NADPH oxidase reveals why phagocytes need proton channels. *Nature.* 2003, 422, 531-534.

[39] DePasquale-Jardieu P, Fraker PJ. Interference in the development of a secondary immune response in mice by zinc deprivation: persistence of effects. *J. Nutr.* 1984, 114, 1762-1769.

[40] Donadelli M, Dalla Pozza E, Costanzo C, Scupoli MT, Scarpa A, Palmieri M. Zinc depletion efficiently inhibits pancreatic cancer cell growth by increasing the ratio of antiproliferative/proliferative genes. *J. Cell Biochem.* 2008, 104, 202-212.

[41] Driessen C, Hirv K, Kirchner H, Rink L. Zinc regulates cytokine induction by superantigens and lipopolysaccharide. *Immunology.* 1995, 84, 272-277.

[42] Driessen C, Hirv K, Rink L, Kirchner H. Induction of Cytokines by Zinc Ions in Human Peripheral Blood Mononuclear Cells and Seperated Monocytes. *Lymphokine Cytokine Res.* 1994, 13, 15-20.

[43] Driessen C, Hirv K, Wellinghausen N, Kirchner H, Rink L. Influence of serum on zinc, toxic shock syndrome toxin-1, and lipopolysaccharide-induced production of IFN-gamma and IL-1 beta by human mononuclear cells. *J. Leukoc. Biol.* 1995, 57, 904-908.

[44] Dubben S, Hönscheid A, Winkler K, Rink L, Haase H. Cellular zinc homeostasis is a regulator in monocyte differentiation of HL-60 cells by 1 alpha,25-dihydroxyvitamin D3. *J. Leukoc. Biol.* 2010, 87, 833-844.

[45] Duchateau J, Delepesse G, Vrijens R, Collet H. Beneficial effects of oral zinc supplementation on the immune response of old people. *Am. J. Med.* 1981, 70, 1001-1004.

[46] Ebara M, Fukuda H, Hatano R, Saisho H, Nagato Y, Suzuki K, Nakajima K, Yukawa M, Kondo F, Nakayama A, Sakurai H. Relationship between copper, zinc and metallothionein in hepatocellular carcinoma and its surrounding liver parenchyma. *J. Hepatol.* 2000, 33, 415-422.

[47] Eby GA, 3rd. Zinc lozenges as cure for the common cold--a review and hypothesis. *Med. Hypotheses.* 2010, 74, 482-492.

[48] Egefjord L, Jensen JL, Bang-Berthelsen CH, Petersen AB, Smidt K, Schmitz O, Karlsen AE, Pociot F, Chimienti F, Rungby J, Magnusson NE. Zinc transporter gene expression is regulated by pro-inflammatory cytokines: a potential role for zinc transporters in beta-cell apoptosis? *BMC Endocr. Disord.* 2009, 9, 7.

[49] El Muayed M, Billings LK, Raja MR, Zhang X, Park PJ, Newman MV, Kaufman DB, O'Halloran TV, Lowe WL, Jr. Acute cytokine-mediated downregulation of the zinc transporter ZnT8 alters pancreatic beta-cell function. *J. Endocrinol.* 2010, 206, 159-169.

[50] Elenkov IJ, Iezzoni DG, Daly A, Harris AG, Chrousos GP. Cytokine dysregulation, inflammation and well-being. *Neuroimmunomodulation.* 2005, 12, 255-269.

[51] Faber C, Gabriel P, Ibs KH, Rink L. Zinc in pharmacological doses suppresses allogeneic reaction without affecting the antigenic response. *Bone Marrow Transplant.* 2004, 33, 1241-1246.

[52] Faure P, Benhamou PY, Perard A, Halimi S, Roussel AM. Lipid peroxidation in insulin-dependent diabetic patients with early retina degenerative lesions: effects of an oral zinc supplementation. *Eur. J. Clin. Nutr.* 1995, 49, 282-288.

[53] Feng P, Li T, Guan Z, Franklin RB, Costello LC. The involvement of Bax in zinc-induced mitochondrial apoptogenesis in malignant prostate cells. *Mol. Cancer.* 2008, 7, 25.

[54] Ferry F, Donner M. In vitro modulation of murine natural killer cytotoxicity by zinc. *Scand. J. Immunol.* 1984, 19, 435-445.

[55] Fischer Walker CL, Lamberti L, Roth D, Black RE. Zinc and Infectious Diseases. In: Rink L, editor. *Zinc in Human Health.* 2011. ISBN: 978-1-60750-815-1.

[56] Fraker PJ, DePasquale-Jardieu P, Zwickl CM, Luecke RW. Regeneration of T-cell helper function in zinc-deficient adult mice. Proc. Natl. Acad. Sci. U. S. A. 1978, 75, 5660-5664.

[57] Fraker PJ, King LE. Reprogramming of the immune system during zinc deficiency. *Annu. Rev. Nutr.* 2004, 24, 277-298.

[58] Fraker PJ, Zwickl CM, Luecke RW. Delayed type hypersensitivity in zinc deficient adult mice: impairment and restoration of responsivity to dinitrofluorobenzene. *J. Nutr.* 1982, 112, 309-313.

[59] Franklin RB, Costello LC. The important role of the apoptotic effects of zinc in the development of cancers. *J. Cell Biochem.* 2009, 106, 750-757.

[60] Freitas M, Porto G, Lima JL, Fernandes E. Zinc activates neutrophils' oxidative burst. *Biometals.* 2010, 23, 31-41.

[61] Gaffen SL, Liu KD. Overview of interleukin-2 function, production and clinical applications. *Cytokine.* 2004, 28, 109-123.

[62] Geissmann F, Manz MG, Jung S, Sieweke MH, Merad M, Ley K. Development of monocytes, macrophages, and dendritic cells. *Science.* 2010, 327, 656-661.

[63] Giacconi R, Cipriano C, Muti E, Costarelli L, Maurizio C, Saba V, Gasparini N, Malavolta M, Mocchegiani E. Novel -209A/G MT2A polymorphism in old patients with type 2 diabetes and atherosclerosis: relationship with inflammation (IL-6) and zinc. *Biogerontology.* 2005, 6, 407-413.

[64] Glimcher LH, Murphy KM. Lineage commitment in the immune system: the T helper lymphocyte grows up. *Genes Dev.* 2000, 14, 1693-1711.

[65] Gonzalez A, Peters U, Lampe JW, White E. Zinc intake from supplements and diet and prostate cancer. *Nutr. Cancer.* 2009, 61, 206-215.

[66] Grivennikov SI, Karin M. Inflammatory cytokines in cancer: tumour necrosis factor and interleukin 6 take the stage. *Ann. Rheum. Dis.* 2011, 70 Suppl 1, i104-108.

[67] Haase H, Maret W. Protein tyrosine phosphatases as targets of the combined insulinomimetic effects of zinc and oxidants. *Biometals.* 2005, 18, 333-338.

[68] Haase H, Mocchegiani E, Rink L. Correlation between zinc status and immune function in the elderly. *Biogerontology.* 2006, 7, 421-428.

[69] Haase H, Ober-Blobaum JL, Engelhardt G, Hebel S, Heit A, Heine H, Rink L. Zinc signals are essential for lipopolysaccharide-induced signal transduction in monocytes. *J. Immunol.* 2008, 181, 6491-6502.

[70] Haase H, Rink L. Functional significance of zinc-related signaling pathways in immune cells. *Annu. Rev. Nutr.* 2009, 29, 133-152.

[71] Haase H, Rink L. The immune system and the impact of zinc during aging. *Immun. Ageing.* 2009, 6, 9.

[72] Hadden JW. Thymic endocrinology. *Ann. N. Y. Acad. Sci.* 1998, 840, 352-358.

[73] Haider BA, Bhutta ZA. The effect of therapeutic zinc supplementation among young children with selected infections: a review of the evidence. *Food Nutr. Bull.* 2009, 30, S41-59.

[74] Hasegawa H, Suzuki K, Nakaji S, Sugawara K. Effects of zinc on the reactive oxygen species generating capacity of human neutrophils and on the serum opsonic activity in vitro. *Luminescence.* 2000, 15, 321-327.

[75] Hashemipour M, Kelishadi R, Shapouri J, Sarrafzadegan N, Amini M, Tavakoli N, Movahedian-Attar A, Mirmoghtadaee P, Poursafa P. Effect of zinc supplementation on insulin resistance and components of the metabolic syndrome in prepubertal obese children. *Hormones.* (Athens), 2009, 8, 279-285.

[76] Ho E. Zinc deficiency, DNA damage and cancer risk. *J. Nutr. Biochem.* 2004, 15, 572-578.

[77] Hönscheid A, Rink L, Haase H. T-lymphocytes: a target for stimulatory and inhibitory effects of zinc ions. *Endocr. Metab. Immune Disord. Drug Targets.* 2009, 9, 132-144.

[78] Hostens K, Pavlovic D, Zambre Y, Ling Z, Van Schravendijk C, Eizirik DL, Pipeleers DG. Exposure of human islets to cytokines can result in

disproportionately elevated proinsulin release. *J. Clin. Invest.* 1999, 104, 67-72.

[79] Huang L, Kirschke CP, Zhang Y. Decreased intracellular zinc in human tumorigenic prostate epithelial cells: a possible role in prostate cancer progression. *Cancer Cell Int.* 2006, 6, 10.

[80] Hujanen ES, Seppa ST, Virtanen K. Polymorphonuclear leukocyte chemotaxis induced by zinc, copper and nickel in vitro. *Biochim. Biophys. Acta.* 1995, 1245, 145-152.

[81] Hulisz D. Efficacy of zinc against common cold viruses: an overview. *J. Am. Pharm. Assoc.* (2003), 2004, 44, 594-603.

[82] Ibs K-H, Rink, L. Zinc. In: Hughes DA, Darlington LG, Bendich A, editors. *Diet and Human Immune Function.* Totowa, N.J.: Humana Press; 2004; p. 241-259.

[83] Iwasaki A, Medzhitov R. Regulation of adaptive immunity by the innate immune system. *Science.* 2010, 327, 291-295.

[84] Jackson JL, Lesho E, Peterson C. Zinc and the common cold: a meta-analysis revisited. *J. Nutr.* 2000, 130, 1512S-1515S.

[85] Jackson MJ. Physiology of Zinc: General Aspects. In: Mills CF, editor. *Zinc in Human Biology.* Berlin Heidelberg: Springer-Verlag; 1989; p. 2-3.

[86] Janeway CA, Jr., Medzhitov R. Innate immune recognition. *Annu. Rev. Immunol.* 2002, 20, 197-216.

[87] Jansen J, Karges W, Rink L. Zinc and diabetes--clinical links and molecular mechanisms. *J. Nutr. Biochem.* 2009, 20, 399-417.

[88] Jarrard DF. Does zinc supplementation increase the risk of prostate cancer? *Arch. Ophthalmol.* 2005, 123, 102-103.

[89] Kagara N, Tanaka N, Noguchi S, Hirano T. Zinc and its transporter ZIP10 are involved in invasive behavior of breast cancer cells. *Cancer Sci.* 2007, 98, 692-697.

[90] Kahmann L, Uciechowski P, Warmuth S, Plumakers B, Gressner AM, Malavolta M, Mocchegiani E, Rink L. Zinc supplementation in the elderly reduces spontaneous inflammatory cytokine release and restores T cell functions. *Rejuvenation Res.* 2008, 11, 227-237.

[91] Kaltenberg J, Plum LM, Ober-Blöbaum JL, Hönscheid A, Rink L, Haase H. Zinc signals promote IL-2-dependent proliferation of T cells. *Eur. J. Immunol.* 2010, 40, 1496-1503.

[92] Karlsen TH, Sommerfelt H, Klomstad S, Andersen PK, Strand TA, Ulvik RJ, Ahren C, Grewal HM. Intestinal and systemic immune responses to an oral cholera toxoid B subunit whole-cell vaccine

administered during zinc supplementation. *Infect. Immun.* 2003, 71, 3909-3913.

[93] Keen CL, Gershwin ME. Zinc deficiency and immune function. *Annu. Rev. Nutr.* 1990, 10, 415-431.

[94] Kim YJ, Kang JH, Yang MP. Zinc increases the phagocytic capacity of canine peripheral blood phagocytes in vitro. *Vet. Res. Commun.* 2009, 33, 251-261.

[95] King LE, Fraker PJ. Flow cytometric analysis of the phenotypic distribution of splenic lymphocytes in zinc-deficient adult mice. *J. Nutr.* 1991, 121, 1433-1438.

[96] King LE, Frentzel JW, Mann JJ, Fraker PJ. Chronic zinc deficiency in mice disrupted T cell lymphopoiesis and erythropoiesis while B cell lymphopoiesis and myelopoiesis were maintained. *J. Am. Coll. Nutr.* 2005, 24, 494-502.

[97] Kinlaw WB, Levine AS, Morley JE, Silvis SE, McClain CJ. Abnormal zinc metabolism in type II diabetes mellitus. *Am. J. Med.* 1983, 75, 273-277.

[98] Kitabayashi C, Fukada T, Kanamoto M, Ohashi W, Hojyo S, Atsumi T, Ueda N, Azuma I, Hirota H, Murakami M, Hirano T. Zinc suppresses Th17 development via inhibition of STAT3 activation. *Int. Immunol.* 2010, 22, 375-386.

[99] Knoell DL, Besecker B. Zinc in Critical Illness and Sepsis. In: Rink L, editor. *Zinc in Human Health.* Amsterdam: IOS Press; 2011. ISBN: 978-1-60750-815-1.

[100] Kojima C, Kawakami A, Takei T, Nitta K, Yoshida M. Angiotensin-converting enzyme inhibitor attenuates monocyte adhesion to vascular endothelium through modulation of intracellular zinc. *J. Pharmacol. Exp. Ther.* 2007, 323, 855-860.

[101] Korn T, Oukka M, Kuchroo V, Bettelli E. Th17 cells: effector T cells with inflammatory properties. *Semin. Immunol.* 2007, 19, 362-371.

[102] Kreft B, Fischer A, Kruger S, Sack K, Kirchner H, Rink L. The impaired immune response to diphtheria vaccination in elderly chronic hemodialysis patients is related to zinc deficiency. *Biogerontology.* 2000, 1, 61-66.

[103] Kröncke KD, Klotz LO. Zinc fingers as biologic redox switches? *Antioxid. Redox Signal.* 2009, 11, 1015-1027.

[104] Kvell K, Cooper EL, Engelmann P, Bovari J, Nemeth P. Blurring borders: innate immunity with adaptive features. *Clin. Dev. Immunol.* 2007, 2007, 83671.

[105] Lassi ZS, Haider BA, Bhutta ZA. Zinc supplementation for the prevention of pneumonia in children aged 2 months to 59 months. *Cochrane Database Syst. Rev.* 2010, CD005978.

[106] Lazzerini M, Ronfani L. Oral zinc for treating diarrhoea in children. *Cochrane Database Syst. Rev.* 2008, CD005436.

[107] Leitzmann MF, Stampfer MJ, Wu K, Colditz GA, Willett WC, Giovannucci EL. Zinc supplement use and risk of prostate cancer. *J. Natl. Cancer Inst.* 2003, 95, 1004-1007.

[108] Levy MM, Fink MP, Marshall JC, Abraham E, Angus D, Cook D, Cohen J, Opal SM, Vincent JL, Ramsay G. 2001 SCCM/ESICM/ACCP/ATS/SIS International Sepsis Definitions Conference. *Crit. Care Med.* 2003, 31, 1250-1256.

[109] Li M, Zhang Y, Liu Z, Bharadwaj U, Wang H, Wang X, Zhang S, Liuzzi JP, Chang SM, Cousins RJ, Fisher WE, Brunicardi FC, Logsdon CD, Chen C, Yao Q. Aberrant expression of zinc transporter ZIP4 (SLC39A4) significantly contributes to human pancreatic cancer pathogenesis and progression. *Proc. Natl. Acad. Sci. U. S. A.* 2007, 104, 18636-18641.

[110] Liang JY, Liu YY, Zou J, Franklin RB, Costello LC, Feng P. Inhibitory effect of zinc on human prostatic carcinoma cell growth. *Prostate.* 1999, 40, 200-207.

[111] Lichten LA, Cousins RJ. Mammalian zinc transporters: nutritional and physiologic regulation. *Annu. Rev. Nutr.* 2009, 29, 153-176.

[112] Lightman A, Brandes JM, Binur N, Drugan A, Zinder O. Use of the serum copper/zinc ratio in the differential diagnosis of ovarian malignancy. *Clin. Chem.* 1986, 32, 101-103.

[113] Liuzzi JP, Lichten LA, Rivera S, Blanchard RK, Aydemir TB, Knutson MD, Ganz T, Cousins RJ. Interleukin-6 regulates the zinc transporter Zip14 in liver and contributes to the hypozincemia of the acute-phase response. *Proc. Natl. Acad. Sci. U. S. A.* 2005, 102, 6843-6848.

[114] Lowe NM, Fekete K, Decsi T. Methods of assessment of zinc status in humans: a systematic review. *Am. J. Clin. Nutr.* 2009, 89, 2040S-2051S.

[115] MacDonald RS. The role of zinc in growth and cell proliferation. *J. Nutr.* 2000, 130, 1500S-1508S.

[116] Mandrup-Poulsen T. The role of interleukin-1 in the pathogenesis of IDDM. *Diabetologia.* 1996, 39, 1005-1029.

[117] Maret W. Metallothionein redox biology in the cytoprotective and cytotoxic functions of zinc. *Exp. Gerontol.* 2008, 43, 363-369.

[118] Maret W, Sandstead HH. Zinc requirements and the risks and benefits of zinc supplementation. *J. Trace Elem. Med. Biol.* 2006, 20, 3-18.

[119] Margalioth EJ, Schenker JG, Chevion M. Copper and zinc levels in normal and malignant tissues. *Cancer.* 1983, 52, 868-872.

[120] Marshall I. Zinc for the common cold. *Cochrane Database Syst. Rev.* 2000, CD001364.

[121] Marshall S. Zinc gluconate and the common cold. Review of randomized controlled trials. *Can. Fam. Physician.* 1998, 44, 1037-1042.

[122] Matsumura M, Nagata M, Nakamura K, Kawai M, Baba T, Yamaki K, Yoshino S. Adjuvant effect of zinc oxide on Th2 but not Th1 immune responses in mice. *Immunopharmacol. Immunotoxicol.* 2010, 32, 56-62.

[123] Maverakis E, Fung MA, Lynch PJ, Draznin M, Michael DJ, Ruben B, Fazel N. Acrodermatitis enteropathica and an overview of zinc metabolism. *J. Am. Acad. Dermatol.* 2007, 56, 116-124.

[124] May JM, Contoreggi CS. The mechanism of the insulin-like effects of ionic zinc. *J. Biol. Chem.* 1982, 257, 4362-4368.

[125] Medzhitov R. Origin and physiological roles of inflammation. *Nature.* 2008, 454, 428-435.

[126] Medzhitov R. Inflammation 2010: new adventures of an old flame. *Cell.* 2010, 140, 771-776.

[127] Milner J. Structures and functions of the tumor suppressor p53. *Pathol. Biol.* (Paris), 1997, 45, 797-803.

[128] Moroni F, Di Paolo ML, Rigo A, Cipriano C, Giacconi R, Recchioni R, Marcheselli F, Malavolta M, Mocchegiani E. Interrelationship among neutrophil efficiency, inflammation, antioxidant activity and zinc pool in very old age. *Biogerontology.* 2005, 6, 271-281.

[129] Mossad SB, Macknin ML, Medendorp SV, Mason P. Zinc gluconate lozenges for treating the common cold. A randomized, double-blind, placebo-controlled study. *Ann. Intern. Med.* 1996, 125, 81-88.

[130] Muzzioli M, Stecconi R, Moresi R, Provinciali M. Zinc improves the development of human CD34+ cell progenitors towards NK cells and increases the expression of GATA-3 transcription factor in young and old ages. *Biogerontology.* 2009, 10, 593-604.

[131] Nolan CJ, Damm P, Prentki M. Type 2 diabetes across generations: from pathophysiology to prevention and management. *Lancet.* 2011, 378, 169-181.

[132] Overbeck S, Rink L, Haase H. Modulating the immune response by oral zinc supplementation: a single approach for multiple diseases. *Arch. Immunol. Ther. Exp. (Warsz)*, 2008, 56, 15-30.

[133] Pan SY, Zhou J, Gibbons L, Morrison H, Wen SW, Ccrerg TC. Antioxidants and breast cancer risk - a population-based case-control study in Canada. *BMC Cancer.* 2011, 11, 372.

[134] Pancer Z, Cooper MD. The evolution of adaptive immunity. *Annu. Rev. Immunol.* 2006, 24, 497-518.

[135] Plum LM, Rink L, Haase H. The essential toxin: impact of zinc on human health. *Int. J. Environ. Res. Public Health.* 2010, 7, 1342-1365.

[136] Porter KG, McMaster D, Elmes ME, Love AH. Anaemia and low serum-copper during zinc therapy. *Lancet.* 1977, 2, 774.

[137] Prasad AS. Effects of zinc deficiency on Th1 and Th2 cytokine shifts. *J. Infect. Dis.* 2000, 182 Suppl 1, S62-68.

[138] Prasad AS. Impact of the discovery of human zinc deficiency on health. *J. Am. Coll. Nutr.* 2009, 28, 257-265.

[139] Prasad AS, Beck FW, Endre L, Handschu W, Kukuruga M, Kumar G. Zinc deficiency affects cell cycle and deoxythymidine kinase gene expression in HUT-78 cells. *J. Lab. Clin. Med.* 1996, 128, 51-60.

[140] Prasad AS, Beck FW, Snell DC, Kucuk O. Zinc in cancer prevention. *Nutr. Cancer.* 2009, 61, 879-887.

[141] Prasad AS, Brewer GJ, Schoomaker EB, Rabbani P. Hypocupremia induced by zinc therapy in adults. *JAMA.* 1978, 240, 2166-2168.

[142] Prasad AS, Fitzgerald JT, Hess JW, Kaplan J, Pelen F, Dardenne M. Zinc deficiency in elderly patients. *Nutrition.* 1993, 9, 218-224.

[143] Prasad AS, Meftah S, Abdallah J, Kaplan J, Brewer GJ, Bach JF, Dardenne M. Serum Thymulin in Human Zinc Deficiency. *J. Clin. Invest.* 1988, 82, 1202-1210.

[144] Provinciali M, Donnini A, Argentati K, Di Stasio G, Bartozzi B, Bernardini G. Reactive oxygen species modulate Zn(2+)-induced apoptosis in cancer cells. *Free Radic. Biol. Med.* 2002, 32, 431-445.

[145] Provinciali M, Montenovo A, Di Stefano G, Colombo M, Daghetta L, Cairati M, Veroni C, Cassino R, Della Torre F, Fabris N. Effect of zinc or zinc plus arginine supplementation on antibody titre and lymphocyte subsets after influenza vaccination in elderly subjects: a randomized controlled trial. *Age Ageing.* 1998, 27, 715-722.

[146] Qadri F, Ahmed T, Wahed MA, Ahmed F, Bhuiyan NA, Rahman AS, Clemens JD, Black RE, Albert MJ. Suppressive effect of zinc on

antibody response to cholera toxin in children given the killed, B subunit-whole cell, oral cholera vaccine. *Vaccine.* 2004, 22, 416-421.

[147] Rajagopalan S, Long EO. Zinc bound to the killer cell-inhibitory receptor modulates the negative signal in human NK cells. *J. Immunol.* 1998, 161, 1299-1305.

[148] Rajagopalan S, Winter CC, Wagtmann N, Long EO. The Ig-related killer cell inhibitory receptor binds zinc and requires zinc for recognition of HLA-C on target cells. *J. Immunol.* 1995, 155, 4143-4146.

[149] Raz I, Karsai D, Katz M. The influence of zinc supplementation on glucose homeostasis in NIDDM. *Diabetes Res.* 1989, 11, 73-79.

[150] Rink L, Gabriel P. Extracellular and immunological actions of zinc. *Biometals.* 2001, 14, 367-383.

[151] Rink L, Haase H. Zinc homeostasis and immunity. *Trends Immunol.* 2007, 28, 1-4.

[152] Rosenkranz E, Prasad AS, Rink L. Immunobiology and Hematology of Zinc. In: Rink L, editor. *Zinc in Human Health.* Amsterdam: IOS Press; 2011. ISBN: 978-1-60750-815-1.

[153] Roth HP, Kirchgessner M. Konzentrationsverlauf an Wachstumshormon, IGF-1, Insulin und C-Peptid in Serum, Hypophyse und Leber von Zn-Mangelratten. *J. Anim. Physiol. a Anim. Nutr.* 1997, 77, 91-101.

[154] Roussel AM, Kerkeni A, Zouari N, Mahjoub S, Matheau JM, Anderson RA. Antioxidant effects of zinc supplementation in Tunisians with type 2 diabetes mellitus. *J. Am. Coll. Nutr.* 2003, 22, 316-321.

[155] Salem SI, Coward WA, Lunn PG, Hudson GJ. Response of the reproductive system of male rats to protein and zinc deficiency during puberty. *Ann. Nutr. Metab.* 1984, 28, 44-51.

[156] Salzman MB, Smith EM, Koo C. Excessive oral zinc supplementation. *J. Pediatr. Hematol. Oncol.* 2002, 24, 582-584.

[157] Samman S, Roberts DC. The effect of zinc supplements on plasma zinc and copper levels and the reported symptoms in healthy volunteers. *Med. J. Aust.* 1987, 146, 246-249.

[158] Sandström B. Diagnosis of zinc deficiency and excess in individuals and populations. *Food Nutr. Bull.* 2001, 22, 133-137.

[159] Santoliquido PM, Southwick HW, Olwin JH. Trace metal levels in cancer of the breast. *Surg. Gynecol. Obstet.* 1976, 142, 65-70.

[160] Sattar N, Scott HR, McMillan DC, Talwar D, O'Reilly DS, Fell GS. Acute-phase reactants and plasma trace element concentrations in non-

small cell lung cancer patients and controls. *Nutr. Cancer.* 1997, 28, 308-312.

[161] Scott BJ, Bradwell AR. Identification of the serum binding proteins for iron, zinc, cadmium, nickel, and calcium. *Clin. Chem.* 1983, 29, 629-633.

[162] Scott DA. Crystalline insulin. *Biochem. J.* 1934, 28, 1592-1602 1591.

[163] Shankar AH, Prasad AS. Zinc and immune function: the biological basis of altered resistance to infection. *Am. J. Clin. Nutr.* 1998, 68, 447S-463S.

[164] Sheikh A, Shamsuzzaman S, Ahmad SM, Nasrin D, Nahar S, Alam MM, Al Tarique A, Begum YA, Qadri SS, Chowdhury MI, Saha A, Larson CP, Qadri F. Zinc influences innate immune responses in children with enterotoxigenic Escherichia coli-induced diarrhea. *J. Nutr.* 2010, 140, 1049-1056.

[165] Shevach EM. CD4+ CD25+ suppressor T cells: more questions than answers. *Nat. Rev. Immunol.* 2002, 2, 389-400.

[166] Simkin PA. Oral zinc sulphate in rheumatoid arthritis. *Lancet.* 1976, 2, 539-542.

[167] Singh M, Das RR. Zinc for the common cold. *Cochrane Database Syst. Rev.* 2011, CD001364.

[168] Staiger H, Machicao F, Stefan N, Tschritter O, Thamer C, Kantartzis K, Schafer SA, Kirchhoff K, Fritsche A, Haring HU. Polymorphisms within novel risk loci for type 2 diabetes determine beta-cell function. *PLoS One.* 2007, 2, e832.

[169] Tanaka Y, Shiozawa S, Morimoto I, Fujita T. Role of zinc in interleukin 2 (IL-2)-mediated T-cell activation. *Scand. J. Immunol.* 1990, 31, 547-552.

[170] Tang X, Shay NF. Zinc has an insulin-like effect on glucose transport mediated by phosphoinositol-3-kinase and Akt in 3T3-L1 fibroblasts and adipocytes. *J. Nutr.* 2001, 131, 1414-1420.

[171] Tarlinton D, Radbruch A, Hiepe F, Dorner T. Plasma cell differentiation and survival. *Curr. Opin. Immunol.* 2008, 20, 162-169.

[172] Tseng SY, Dustin ML. T-cell activation: a multidimensional signaling network. *Curr. Opin. Cell Biol.* 2002, 14, 575-580.

[173] Tsou TC, Yeh SC, Tsai FY, Lin HJ, Cheng TJ, Chao HR, Tai LA. Zinc oxide particles induce inflammatory responses in vascular endothelial cells via NF-kappaB signaling. *J. Hazard Mater.* 2010, 183, 182-188.

[174] Turk S, Bozfakioglu S, Ecder ST, Kahraman T, Gurel N, Erkoc R, Aysuna N, Turkmen A, Bekiroglu N, Ark E. Effects of zinc

supplementation on the immune system and on antibody response to multivalent influenza vaccine in hemodialysis patients. *Int. J. Artif. Organs.* 1998, 21, 274-278.

[175] Turvey SE, Broide DH. Innate immunity. *J. Allergy Clin. Immunol.* 2010, 125, S24-32.

[176] Valentiner-Branth P, Shrestha PS, Chandyo RK, Mathisen M, Basnet S, Bhandari N, Adhikari RK, Sommerfelt H, Strand TA. A randomized controlled trial of the effect of zinc as adjuvant therapy in children 2-35 mo of age with severe or nonsevere pneumonia in Bhaktapur, Nepal. *Am. J. Clin. Nutr.* 2010, 91, 1667-1674.

[177] Vallee BL, Falchuk KH. The biochemical basis of zinc physiology. *Physiol. Rev. 1993*, 73, 79-118.

[178] Vaziri-Sani F, Oak S, Radtke J, Lernmark K, Lynch K, Agardh CD, Cilio CM, Lethagen AL, Ortqvist E, Landin-Olsson M, Torn C, Hampe CS. ZnT8 autoantibody titers in type 1 diabetes patients decline rapidly after clinical onset. *Autoimmunity.* 2010, 43, 598-606.

[179] Verhaegh GW, Parat MO, Richard MJ, Hainaut P. Modulation of p53 protein conformation and DNA-binding activity by intracellular chelation of zinc. *Mol. Carcinog.* 1998, 21, 205-214.

[180] Vivier E, Raulet DH, Moretta A, Caligiuri MA, Zitvogel L, Lanier LL, Yokoyama WM, Ugolini S. Innate or adaptive immunity? The example of natural killer cells. *Science.* 2011, 331, 44-49.

[181] von Bülow V, Rink L, Haase H. Zinc-mediated inhibition of cyclic nucleotide phosphodiesterase activity and expression suppresses TNF-alpha and IL-1 beta production in monocytes by elevation of guanosine 3',5'-cyclic monophosphate. *J. Immunol.* 2005, 175, 4697-4705.

[182] Waeytens A, De Vos M, Laukens D. Evidence for a potential role of metallothioneins in inflammatory bowel diseases. *Mediators Inflamm.* 2009, 2009, 729172.

[183] Wapnir RA. Zinc deficiency, malnutrition and the gastrointestinal tract. *J. Nutr.* 2000, 130, 1388S-1392S.

[184] Wellenreuther G, Cianci M, Tucoulou R, Meyer-Klaucke W, Haase H. The ligand environment of zinc stored in vesicles. *Biochem. Biophys. Res. Commun.* 2009, 380, 198-203.

[185] Weninger W, Manjunath N, von Andrian UH. Migration and differentiation of CD8+ T cells. *Immunol. Rev.* 2002, 186, 221-233.

[186] Wenzlau JM, Juhl K, Yu L, Moua O, Sarkar SA, Gottlieb P, Rewers M, Eisenbarth GS, Jensen J, Davidson HW, Hutton JC. The cation efflux

transporter ZnT8 (Slc30A8) is a major autoantigen in human type 1 diabetes. *Proc. Natl. Acad. Sci. U. S. A.* 2007, 104, 17040-17045.

[187] Wessels I, Jansen J, Rink L, Uciechowski P. Neutrophils: Phagocytes and Gate Keepers of Inflammation. In: DeFranco JE, editor. Lifespan, *Functions and Roles in Disease:* Nova Science Publishers, Inc.; 2010; p. 121-183.

[188] Weston WL, Huff JC, Humbert JR, Hambidge KM, Neldner KH, Walravens PA. Zinc correction of defective chemotaxis in acrodermatitis enteropathica. *Arch. Dermatol.* 1977, 113, 422-425.

[189] WHO. *The World Health Report.* 2002. 2002.

[190] WHO/UNICEF. Joint Statement: Clinical Mangement of Acute Diarrhoea. 2004.

[191] Wong HR, Shanley TP, Sakthivel B, Cvijanovich N, Lin R, Allen GL, Thomas NJ, Doctor A, Kalyanaraman M, Tofil NM, Penfil S, Monaco M, Tagavilla MA, Odoms K, Dunsmore K, Barnes M, Aronow BJ. Genome-level expression profiles in pediatric septic shock indicate a role for altered zinc homeostasis in poor outcome. *Physiol Genomics.* 2007, 30, 146-155.

[192] Wu FY, Wu CW. Zinc in DNA replication and transcription. *Annu. Rev. Nutr.* 1987, 7, 251-272.

[193] Yeh SC, Tsai FY, Chao HR, Tsou TC. Zinc ions induce inflammatory responses in vascular endothelial cells. *Bull. Environ. Contam. Toxicol.* 2011, 87, 113-116.

[194] Zampelas A. Zinc supplementation: Another myth or we are heading towards a new era in the treatment of diabetes? *Atherosclerosis.* 2011.

[195] Zipfel PF. Complement and immune defense: from innate immunity to human diseases. *Immunol. Lett.* 2009, 126, 1-7.

In: Zinc: Characteristics, Uses and Benefits ISBN: 978-1-61942-523-1
Editor: Masayoshi Yamaguchi

Chapter 7

ZINC AND BONE HOMEOSTASIS: ZINC SUPPLEMENTATION PREVENTS OSTEOPOROSIS

Masayoshi Yamaguchi*
Division of Endocrinology and Metabolism and Lipids,
Department of Medicine,
Emory University School of Medicine,
Atlanta, Georgia 30322-0001, U.S.A.

ABSTRACT

Zinc is known as an essential nutritional factor in the growth of the human and animals. Bone growth retardation is a common finding in various conditions associated with dietary zinc deficiency. Bone zinc content has been shown to decrease in aging, skeletal unloading and postmenopausal conditions, suggesting its role in bone disorder. Zinc has been demonstrated to have a stimulatory effect on osteoblastic bone formation and mineralization; the metal directly activates aminoacyl-tRNA synthetase, a rate-limiting enzyme at translational process of protein synthesis, in the cells, and it stimulates cellular protein synthesis. Zinc has been shown to stimulate gene expression of the transcription factors runt-related transcription factor 2 (Runx 2) that is related to differentiation into osteoblastic cells. Moreover, zinc has been shown to inhibit osteoclastic bone resorption due to inhibiting osteoclast-like cell

* E-mail: yamamasa1155@yahoo.co.jp

formation from bone marrow cells and stimulating apoptotic cell death of mature osteoclasts. Zinc has a suppressive effect on the receptor activator of nuclear factor (NF)-κB ligand (RANKL)-induced osteoclastogenesis. Zinc transporter has been shown to express in osteoblastic cells and osteoclastic cells. Zinc protein is invloved in transcription. The intake of dietary zinc causes an increase in bone mass. β-Alanyl-L-histidinato zinc (AHZ) is a zinc compound, in which zinc is chelated to β-alanyl-L-histidine. The stimulatory effect of AHZ on bone formation is more intensive than that of zinc sulfate. Zinc acexamate has also been shown to have a potent-anabolic effect on bone. The oral administration of AHZ or zinc acexamate has the restorative effect on bone loss under various pathophysiological conditions including aging, skeletal unloading, aluminium bone toxicity, calcium- and vitamin D-deficiency, adjuvant arthritis, estrogen deficiency, diabetes, and fracture healing. Zinc compounds may be designed as new supplymentation factor in the prevention and therapy of osteoporosis.

Keywords: zinc, bone formation, bone resorption, osteoporosis.

INTRODUCTION

Zinc, a nutritional trace element, is essential for the growth of human and animals [1-3]. Zinc is required for the growth, development, and maintenance of healthy bones. Bone growth retardation is a common finding in various conditions associated with zinc deficiency [4-8]. In Iranian schoolboys at year 1961, zinc supplementation was firstly found to restore both skeletal growth and maturation [9]. Zinc deficiency is associated with many kinds of skeletal abnormalities in fetal and postnatal development. Zinc may play a physiologically important role in bone homeostasis.

Skeleton contains a large proportion of the total body burden of zinc [10]. Bone zinc has been shown to be concentrated in the layer of osteoid prior to calcification [11]. Zinc can be mobilized by conditions such as calcium deficiency in the pregnant rats [12]. Early mobilization of maternal bone provided sufficient zinc prevents fetal malformation, and most fetal zinc is derived from maternal muscle mobilized later in the calcium-deficient pregnancy [13].

Bone appears to act as a zinc sink. Zinc is released during skeletal breakdown, and it is mostly reincorporated into the skeleton [14, 15]. In human, the vertebral calcium/zinc is inversely related to age, suggesting that skeletal zinc is conserved more than calcium in later life [16]. Zinc also occurs in the mineral component of bone, probably in hydroxyapatite [14, 17]; it may be complex with fluoride; and both zinc and the zinc-fluoride complex may improve the crystallinity of apatite [18].

Osteoporotic patients have been shown to have lower levels of skeletal zinc than control [19]. In postmenopausal women, urinary zinc has been suggested as a marker of bone resorption, since women with osteoporosis excrete over than 800 μg zinc/g creatinine in urine [10]. The supplements of trace minerals with or without calcium in postmenopausal women have been shown to have beneficial effects on bone density [20].

Zinc may play a physiologically important role in bone. The mechanism of action of zinc on osteoblastic and osteoclastic cells that regulate bone homeostasis, however, has been poorly defined. In recent years, there is growing evidence that zinc has an important role in the regulation of bone homeostasis. Zinc has been demonstrated to stimulate osteoblastic bone formation and to inhibit osteoclastic bone resorption, thereby increasing bone mass. Supplementation of nutritional zinc has been shown to have preventive and therapeutic effects on bone loss that is induced in various bone disorders.

This chapter has been written to outline the recent advances that have been made concerning the role of zinc in the regulation of bone homeostasis and in the prevention and therapy of osteoporosis.

ZINC AND BONE HOMEOSTASIS

Bone Homeostasis

Bone contains over 98% of total body calcium. Bone homeostasis is regulated with the functions of osteoblasts and osteoclasts, which are major cells in bone tissue [21-23]. In the physiological process of bone turnover, a resorptive stimulus firstly triggers recruitment of osteoclasts to a site on the bone surface. Osteoclasts, which develop from hematopoietic progenitors, are recruited to the site and excavate the calcified matrix. Then, the cavity is refilled by osteoblasts via a process that occurs in three distinct phases: initiation, progression and termination. During the initiation phase, a team of osteoblasts arising from local mesenchymal stem cells assembles at the bottom of the cavity and bone formation begins.

As bone formation progresses, some osteoblasts are entombed within the matrix as osteocytes but the majority dies by apoptosis. Bone formation terminates when the cavity has been refilled, at which time the few osteoblasts that remain become the flat lining cells that cover the quiescent surfaces of bone. Once formed, few osteocytes die. Their viability is likely maintained by physiological levels of mechanical stimulation. When mechanical forces are reduced, for example in weightlessness, osteocytes die by apoptosis. This event appears to act as a beacon for osteoclast recruitment and generation of a new basic multicellular unit, which in turn replaces the old bone containing dead osteocytes with new bone containing viable osteocytes.

The process of bone remodeling is regulated with respect to the interactions along the remodeling sequence by systemic influences (hormones), the stress action on trabecular and cortical systems with physical activity and weight bearing, various growth factors produced from the bone cells which act locally on the cells, or other factors that come from nearby cells present in the marrow tissues.

Bone acts as major storage site for growth factors [23]. Growth factors, which are produced by osteoblasts, diffuse into newly deposited osteoid and are stored in the bone matrix including insulin-like growth factors (IGF- I and II), transforming growth factor-β1 (TGF-β1), or platelet-derived growth factor (PDGF). These bone-derived factors, which can be liberated during subsequent periods of bone resorption, act in an autocrine, paracrine, or delayed paracrine fashion in the local microenvironment of the bone surface.

Zinc Has an Anabolic Effect on Bone

Zinc has an anabolic effect on bone metabolism *in vivo* and *in vitro*. The administration of zinc sulfate (5 and 10 mg Zn/kg body weight) for 3 days produced dose-dependent increases in the contents of zinc, deoxyribonucleic acid (DNA), collagen and calcium, and the activity of alkaline phosphatase in the femoral diaphysis (cortical bone) of weanling rats [24]. Alkaline phosphatase is related to bone mineralization in osteoblasts. Collagen is a main bone matrix protein that is produced in osteoblasts. DNA content in bone tissues is a marker of the number of bone cells, including osteoblasts, osetoclasts, and osteocytes. Zinc accumulated in the bone tissues may firstly cause the activation of alkaline phosphatase and the stimulation of collagen synthesis in osteoblasts, which are involved in bone mineralization and calcification.

The anbolic effect on bone is not seen in other essential trace metals [24-26]. The oral administration of chromium (III), cobalt, copper, manganese, nickel, and selenium (15.3 μmol as metal/kg) for 3 days caused a significant increase in the diaphyseal alkaline phosphatase activity and DNA content, while the doses of 153 μmol/kg clearly decreased the enzyme activity. The dose of germanium (306 μmol /kg) caused a significant decrease in the diaphyseal alkaline phosphatase activity and DNA content. The supplementation with copper in adequate copper nutrition does not have anabolic effects on bone components [27]. The oral administration of zinc with the dose of 3,060 μmol /kg produced an increases in both alkaline phosphatase activity and DNA content in the diaphyseal tissues. Zinc has the lowest toxicity for bone metabolism as compared with other trace metals.

Zinc has a stimulatory effect on bone formation in tissue culture system *in vitro*. Calvaria were removed from weanling rats, and it cultured for a period up to 96 h in Dulbecco's Modified Eagle Medium [28]. Zinc uptake by bone tissues was significantly increased in culture with zinc. Bone calcium content was significantly increased in the presence of 10^{-4} M zinc. Physiological zinc concentration in the serum was in the range of 10^{-4} to 4 x10^{-4} M. Bone alkaline phosphatase and ATPase activities were increased in the presence of zinc (10^{-6} to 10^{-3} M), whereas it did not change significantly the activities of pyrophosphatase, acid phosphatase, and β-N-acetylglucosaminidase. Bone collagen content was raised in the presence of zinc. The effect of zinc in increasing bone alkaline phosphatase activity and collagen content has been shown to prevent after the administration of an inhibitor of protein synthesis, suggesting that bone protein synthesis is a necessary component of this response.

The first steps in the biosynthesis of protein involve the enzymatic activation of the amino acids with adenosine triphosphate, followed by the transfer of the amino acids to amino acid-specific ribonucleic acids (RNA). The presence of zinc in the culture medium has been found to induce a significant increase in the incoporation of [^{3}H]-leucine into the acid-insoluble residues of bone tissue [29]. The activity of [^{3}H] leucyl-tRNA synthetase in the 105,000 *g* supernatant fraction (cytosol) of the bone homogenate was increased about twofold after culture with zinc [30]. Zinc has a direct stimulatory effect on protein synthesis at the translational level in bone cells *in vitro*.

Endogenous zinc in the bone tissues has also been shown to play an essential role on bone protein synthesis in the culture [31]. When the calvaria were cultured for 24 h in a medium containing dipicolinate (10^{-6} to 10^{-3} M), a chelator of zinc, bone zinc content was decreased after culture with dipicolinate [31]. In this case, bone alkaline phoshatase activity was decreased by about 40% of untreated bone enzyme activity. The decreased alkaline phosphatase activity was markedly increased in the presence of 10^{-4} M zinc (about 2.5-fold of control value). The effect of zinc was completely blocked after culture with protein synthesis inhibitor. Thus, bone endogenous zinc plays a role in the stimulation of protein synthesis at the translational process in bone cells. Zinc has been shown to increase many constitutive proteins in bone tissues *in vitro* [32].

Zinc has been shown to enhance the anabolic effect of vitamin D_3 on bone components in the femur of weanling rats *in vivo* [33]. The administration of vitamin D_3 (10 μg/kg) or zinc (10 mg Zn/kg) produced a significant increase in bone alkaline phosphatase activity and DNA content. The increase in alkaline phosphatase activity was additionally enhanced by the simultaneous administration of vitamin D_3 and zinc. The increase in bone DNA content was markedly (about fourfold) enhanced by both treatments. These increases were prevented after the administration of cycloheximide, an inhibitor of protein synthesis [33, 34]. The effect of 1,25-dihydroxyvitamin D_3 (0.5 and 1.5 μg/kg) in increasing bone alkaline phosphatase activity and DNA content was synergistically enhanced after the simultaneous treatment with zinc [34]. The receptors for 1,25-dihydroxyvitamin D_3 are shown to have two zinc fingers at the sit of interaction with DNA [35]. Zinc may potentiate the interaction of the 1,25-dihydroxyvitamin D_3-receptor with DNA at that site. These findings suggest that the combination of vitamin D_3 and zinc has a synergistic effect on the stimulation of bone growth and mineralization in rats *in vivo*.

Zinc has also been shown to modulate hormonal effect on bone formation and calcification in bone tissue culture *in vitro*. The presence of 1,25-dihydroxyvitamin D_3 produced a significant increase in alkaline phosphatase activity, DNA, and calcium contents in tissue culture with rat calvaria [36]. The anabolic effect of estrogen (17β-estradiol) on bone components in tissue culture was significantly enhanced after culture with zinc sulfate [37]. The effect was abolished with cycloheximide. Zinc modulates anabolic effect of 1,25-dihydroxyvitamin D_3 or estrogen on bone metabolism *in vitro*.□Zinc has been reported to increase the activity of 1,25-dihydroxyvitamin D_3-dependent promoters in osteoblastic cells [38].

Zinc Stimulates Osteoblastic Bone Formation

The cellular mechanism of zinc action in stimulating bone formation is demonstrated in osteobalstic MC3T3-E1 cells. The proliferation of osteoblastic cells was stimulated after culture with zinc compound (zinc sulfate or zinc-chelating dipeptide) with or without fetal bovine serum [39].

Culture with zinc increased DNA content in the cells. Zinc-induced increase in cell proliferation and DNA content was prevented after culture with cycloheximide. The effect of zinc in stimulating cell proliferation may be mediated through protein synthesis [39]. Zinc has been shown to stimulate cell differentiation of osteoblastic cells [40]. Culture with zinc (10^{-7} to 10^{-5} M) produced a remarkable increase in alkaline phosphatase activity and protein concentration in osteoblastic cells. These increases were also seen with the prolonged cultivation (12-21 days). The effect of zinc was completely abolished after culture with cycloheximide.

Protein components in osteoblastic MC3T3-E1 cells cultured with zinc have been characterized [41]. The homogenate of cells was analyzed with SDS-polyacrylamide gel electrophoresis. Culture with zinc caused an appreciable increase in many protein components in the cells. Especially, the 66 kDa and 44 kDa proteins, which are the major components from control cells, were clearly found to increase in the presence of zinc compound. Moreover, the concentrations of osteocalcin, IGF-I, or TGF-β1 in the culture medium secreted from osteoblastic cells were markedly increased after culture with zinc. Zinc has been found to increase production of bone growth factors and bone matrix protein, which are involved in the stimulation of bone formation and mineralization in osteoblastic cells [41].

The effect of IGF-I in increasing protein concentration, DNA content, and cell number in the cells was markedly enhanced in the presence of zinc [42]. The cellular alkaline phosphatase activity was synergistically increased in the presence of both IGF-I and zinc. Such an effect was not seen after culture with both insulin and zinc. The enhancement of IGF-I's anabolic effect after culture with zinc has been shown to mediate through signaling pathway of protein kinase C and protein phosphatase in osteoblastic cells.

Culture with zinc sulfate has also been shown to increase protein tyrosine phosphatase activity in the cells [43]. The effect of IGF-I in increasing the enzyme activity was enhanced after culture with zinc [43]. Such an effect was not seen with parathyroid hormone. Zinc modulates the anabolic effect of IGF-I on protein tyrosine phosphatase activity and cell proliferation. Likewise, the cell proliferative effect of estrogen in osteoblastic cells has also been enhanced after culture with zinc compound.

The clarification of a molecular mechanism by which zinc stimulates bone protein synthesis has been attempted. Aminoacyl-tRNA synthetase is an enzyme that synthesizes aminoacyl-tRNA. Zinc has been found to activate [^{3}H]-leucyl-tRNA synthetase in the homogenate of osteoblastic cells [44].

Zinc has also been demonstrated to stimulate DNA synthesis in the homogenate of osteoblastic cells *in vitro* [45]. The culture with zinc compound clearly stimulated DNA synthesis in the homogenate of osteoblastic cells, when it was estimated with the incorporation of [^{3}H] deoxythimidine 5'-triphosphate into the DNA in the homogenate. The effect of zinc was completely abolished after culture with cycloheximide, suggesting that the action of zinc is based on newly synthesized protein components. The effect of zinc in stimulating DNA synthesis may mainly result from protein synthesis. Also, it is possible that zinc has an effect on the process of DNA synthesis that is involved in the direct activation of DNA polymerase that is a zinc enzyme.

Zinc has been shown to stimulate the expression of transcription factors runt-related transcription factor 2 (Runx2) mRNA, a transcription factor, which is related to the differentiation to pre-osteoblastic cells [46].

In conclusion, zinc stimulates cell proliferation, cell differentiation, and mineralization in osteoblasts, thereby promoting bone formation. Molecular mechanism of zinc action may be to stimulate gene expression of various proteins including Runx2/Cbfa 1 (Core binding factor alpha1), type I collagen, alkaline phosphatase, and osteocalcin in the cells. Also, zinc increases production of IGF-I and TGF-β1 in the cells. Moreover, zinc enhances protein synthesis due to activating aminoacyl-tRNA synthetase, a rate-limiting enzyme at translational process, in osteoblastic cells. Thus, zinc has a potent stimulatory effect on osteoblasitc bone formation. Interestingly, the effect of IGF-I or estrogen on cell proliferation has been shown to enhance by zinc. The molecular mechanism of zinc action on cell nuclear events remains to be elucidated.

Zinc Suppresses Osteoclastic Bone Resorption

Zinc has an inhibitory effect on bone resorption [47]. Calvaria from weanling rats were cultured for periods of up to 48 h in a medium containing various bone-resorbing factors [parathytoid hormone (PTH), prostaglandin E_2 (PGE_2), interleukin-1α (IL-1α), and lipopolysacharide]. These factors caused a significant decrease in bone calcium content. The decrease in bone calcium content was completely inhibited after culture with zinc compound [48]. Also, zinc compound completely inhibited the PTH- or IL-1α-induced increases in medium glucose consumpion and lactic acid production by bone tissues [47]. The inhibitory effect of zinc compound on bone resorption was not seen after culture with dipicolinate, a chelator of zinc. Thus, zinc has been shown to have an inhibitory effect on bone resorption in tissue culture system *in vitro*.

PGE_2 is secreted from osteoblasts. PTH or IL-1α caused a remarkable elevation of PGE_2 production in osteoblasts *in vitro* [48]. Culture with zinc compound did not have an effect on PGE_2 production in osteoblasts [48], indicating that the effect of zinc may not mediated through suppression of PGE_2 production in osteoblasts.

Bone-resorbing cells, osteoclasts, are formed by differentiation of bone marrow cells. Zinc has an inhibitory effect on osteoclast-like cell formation in mouse marrow culture *in vitro* [49]. The bone marrow cells were cultured for 7 days in a medium containing bone-resorbing agent. Osteoclast-like cell formation was estimated staining for tartrate-resistant acid phosphatase (TRACP), a marker enzyme of osteoclasts. The presence of 1,25-dihydroxyvitamin D_3, PTH, IL-1α, or PGE_2 induced a remarkable increase in osteoclast-like multinucleated cells. These increases were inhibited after culture with zinc (10^{-8} to 10^{-6} M). The inhibitory effect of zinc compound was equal in comparison with the effect of other anti-bone resorbing agents (calcitonin, 17β-estradiol, or acetazolamide) on osteoclast-like cell formation in mouse marrow culture. When osteoclast isolated from rat femoral-diaphyseal tissues were cultured for 24 h in the presence of zinc, the metal did not have an effect on lysosomal enzyme activity (acid phosphatase or β-glucuronidase) in the cells [50]. Culture with zinc caused apoptotic cell death of mature osteoclast-like cells isolated in rat femoral tissues [50]. Zinc has been shown to have suppressive effects on osteoclastogenesis and osteoclastic cell death that are generated with differentiation of bone marrow cells.

The mechanism of zinc action in inhibiting the PTH-induced osteoclast-like cell formation in mouse marrow culture system *in vitro* has been shown [51]. The effect of zinc in inhibiting PTH-induced osteoclast-like cell formation was clearly seen in the presence or absence of theophylline. However, zinc did not inhibit the stimulatory effect of dibutyryl cyclic AMP on osteoclast-like cell formation. The stimulatory effect of PTH on osteoclast-like cell formation was clearly weakened (about 50%) in the presence of EGTA or dibucaine, a regulatory factor of intracellular Ca^{2+} signaling. Phorbol 12-myristate 13-acetate (PMA), a protein kinase C activator, stimulated osteoclast-like cell formation. The effect of PMA was inhibited after culture with zinc. However, the inhibitory effect of zinc was not seen in the presence of both PTH and PMA. These findings support the view that zinc inhibits PTH-stimulated osteoclast-like cell formation mediated through the Ca^{2+}-dependent activation of protein kinase C.

Receptor activator of NF-κB ligand (RANKL) plays a pivotal role in development of osteoclasts from pre-osteoclasts [52, 53]. RANKL is secreted from osteoblasts. RANKL is a member of the tumor necrosis factor (TNF) superfamily, which was originally identified as T-cell-derived immunomodulatory cytokines [54]. RANKL is expressed in activated T cells and promotes the survival of dendritic cells by binding to its receptor RANK. RANKL/RANK pathway is essential for osteoclast differentiation [52, 53]. RANKL is expressed in osteoblastic cells and bone marrow stromal cells in response to osteotropic factors. The combined treatment of hematopoietic cells with macrophage-colony stimulating factor (M-CSF) and the soluble from of RANKL (sRANKL) induces osteoclast differentiation *in vitro* [55]. The effect of RANKL was completely abrogated by adding a natural antagonist of RANKL, osteoprotegerin (OPG) [56], which is produced in osteoblastic cells. TNF receptor-associated factor (TRAF) family proteins are adaptor molecules that mediate intracellular signaling of various cytokine receptors including the TNF receptor superfamily and Toll/interleukin receptor (IL-1R) family [57]. TRAF6 binds to the membrane-proximal region of RANK and IL-1R-associated kinase, and is critically involved in the intracellular signal transduction including NF-κB and mitogen-activated protein kinase (MAPK) activation.

Zinc has been shown to have an inhibitory effect on RANKL-induced osteoclast-like cell formation in mouse marrow culture in the presence of M-CSF [58]. Zinc also inhibited TNFα-induced osteoclastogenesis [57]. TNFα is an autocrine factor in osteoclasts, promoting their differentiation, and mediates, at least in part, RANKL's induction of osteoclastogenesis [59]. The inhibitory effect of zinc on osteoclastogenesis may be partly involved in the suppressive effect on RANKL stimulation. In addition, zinc may inhibit signaling pathway that is related to RANKL stimulation in pre-osteoclasts.

The effect of zinc in inhibiting RANKL-induced osteoclastogenesis has been shown to abolish after culture with inhibitors of protein synthesis or trascription activity in mouse marrow cultures [58]. This finding suggests that the inhibitory effect of zinc on osteoclastogenesis is partly resulted from newly synthesized protein components that are involved in gene expression in mouse marrow culture.

Culture with zinc has been shown to have stimulatory effect on the expression of OPG mRNA in osteoblastic cells [46]. It is speculated that zinc stimulates the expression of RANKL inhibitor (including OPG) in osteoblasts, and that the metal induces a factor that can suppress osteoclast development in pre-osteoclasts.

Role of Zinc in Bone Growth

Zinc is an essential element for the growth of human and animals. Nutritional zinc is required for the growth, development, and maintenance of healthy bones. Bone growth retardation is a common finding in various conditions associated with zinc deficiency [4-8]. In fact, zinc has been demonstrated to stimulate bone growth in newborn rats supplied with lactation by maternal rats [60]. Newborn rats were killed between 1 and 35 days after birth. Increasing age caused a significant increase in zinc and calcium contents and alkaline phosphatase activity in the femoral-diaphyseal and –metaphyseal tissues. The oral administration of zinc sulfate (20 mg Zn/kg body weight) with four times at 24-h intervals to maternal rats from 1 day after birth induced a significant increase in zinc, alkaline phosphatase activity, DNA and calcium contents in the femoral-diaphyseal and –metaphyseal tissues of newborn rats compared with those at 7 or 14 days old [60], indicating that the increase in bone components results from lactation with zinc-containing milk of maternal rats. Zinc plays a physiological role in the development of bone growth in newborn rats [60].

Endogenous zinc may be important in protein synthesis in the femoral-diaphyseal and –metaphyseal tissues of newborn rats [61]. Bone tissues were obtained at 1, 7, 14, 21, and 28 days after birth. Many protein molecules were found to be present in the bone tissues using SDS-polyacrylamide gel electrophoresis analysis. Bone protein synthesis activity was enhanced with increasing age, and reached a plateau 21 days after birth. Protein synthesis in the diaphyseal and metaphyseal tissues obtained from 7- or 14-day-old rats was significantly decreased after culture with a chelator of zinc ion [61]. This decrease was completely blocked after culture with addition of zinc. Zinc-induced increase in bone protein synthesis was completely prevented after culture with cycloheximide or actinomycin D. Thus, bone protein synthesis has been shown to enhance with increasing age of newborn rats, and endogenous zinc in bone tissues plays a physiologic role in the enhancement of protein synthesis in bone growth [61].

Zinc has been found to stimulates the productions of IGF-I and TGF-β1 in the bone tissues of newborn rats [62]. Femoral-diaphyseal and –metaphyseal tissues were obtained at 1, 7, 14, 21, and 28 days after birth of newborn rats, and cultured for 24 h in serum-free medium. Protein concentration in the culture medium was significantly increased when cultured bone tissues from newborn rats with increasing age. Medium IGF-I or TGF-β1 concentrations were gradually reduced with increasing age after birth. The presence of zinc sulfate (10^{-5} or 10^{-4} M) caused a significant increase in protein, IGF-I, or TGF-β1 concentrations in the culture medium. The expression of IGF-I or TGF-β1 mRNAs were significantly increased after culture with zinc (10^{-4} M). Zinc has been shown to have stimulatory effects on IGF-I or TGF-β1 production in the femoral tissues with bone growth of newborn rats [62].

Bone growth factors are physiologically important in the stimulation of bone growth of newborn rats [63, 64]. Zinc has a physiological role in the stimulation of bone growth in collaboration with IGF-I or TGF-β1 [65]. Culture with zinc has been shown to enhance the effect of IGF-I in increasing the culture medium protein concentration, alkaline phosphatase activity or DNA content in the femoral-diaphyseal and –metaphyseal tissues obtained from newborn rats. Such an effect of zinc was not seen in the case of TGF-β1. Zinc may partly act on protein tyrosine kinase and protein tyrosine phosphatase that are related to signaling mechanism of IGF-I in the cells [43].

Zinc has been shown to stimulate *in vitro* DNA synthesis activity in the femoral-diaphyseal and –metaphyseal tissues obtained from newborn rats [66]. An increase in bone DNA synthesis activity was associated with bone growth of newborn rats with aging (1- 21 days). Bone DNA synthesis activity has been shown to reduce after culture with zinc chelator in the reaction mixture of assay system for DNA synthesis [66], indicating that bone endogenous zinc plays a role in the enhancement of bone DNA synthesis associated with bone growth of newborn rats. Also, zinc has a direct stimulatory effect on DNA synthesis in the femoral tissues obtained from newborn rats [66]. DNA polymerase, which is related to DNA synthesis, is a zinc enzyme. It is possible that zinc partly stimulates DNA synthesis due to activating DNA polymerase in osteoblastic cells of bone tissues.

Bone DNA synthesis activity has been significantly increased after culture with IGF-I or TGF-β1 [66]. The effect of IGF-I in increasing bone DNA synthesis activity was significantly enhanced in the presence of zinc. Such an effect was not seen in the case of TGF-β1. The effect of zinc, IGF-I, or zinc plus IGF-I in increasing bone DNA synthesis activity was completely prevented after culture with an inhibitor of mitogen-activated protein (MAPK) kinase. Zinc has a stimulatory effect on bone DNA synthesis in newborn rats. Zinc has been demonstrated to increase proliferation of osteoblastic MC3T3-E1 cells *in vitro* [39]. Presumably, zinc activates MAPK kinase, and it has a stimulatory effect on the proliferation of osteoblastic cells in the bone tissues of newborn rats [64]. Zinc deficiency has been reported to induce a decrease in serum IGF-I level and a retardation of skeletal growth in young rats [5]. Zinc may play an important role in the stimulation of bone growth in collaboration with IGF-I in newborn rats.

The cellular and molecular mechanism by which zinc stimulates bone growth in newborn rats has been summarized in Figure 1.

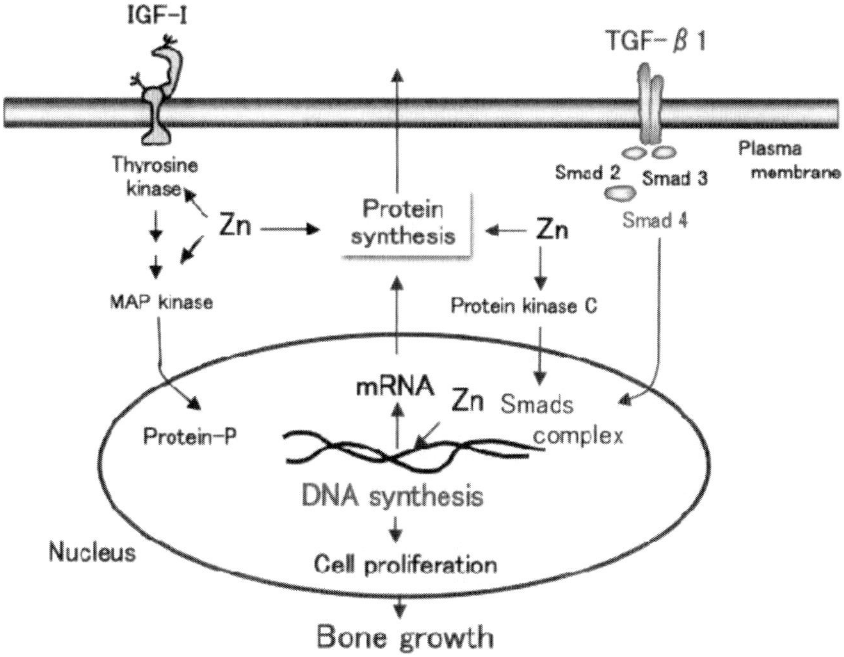

Figure 1. Role of zinc in the stimulation of bone growth. Zinc increases protein synthesis at translational process due to activating aminoacyl-tRNA synthetase in osteoblastic cells. Zinc activates MAPK kinase or protein kinase C that is related to signaling in gene expression, and it may directly enhance gene expression. Zinc also stimulates DNA synthesis in the cells. Moreover, zinc increases production of IGF-I and TGF-β1 which act osteoblastic cells, and the metal modulates the stimulatory effects of IGF-I and TGF-β1 on cell proliferation and their related cell functions in the cells.

ROLE OF ZINC IN FRACTURE HEALING

Fracture repair can be considered as a biologically optimal process resulting in the restoration of injured skeletal tissue to a state of normal structure and function [67]. The mechanism by which bone fracture heals is complex. Fracture healing can be envisioned as involving five distinguishable processes, including the immediate response to injury, intramembranous bone formation, chondrogenesis, endochondral bone formation leading to the reestablishment of load bearing function, and bone remodeling [67, 68]. It is recognized that these processes may occur simultaneously during fracture repair. During fracture healing, a number of growth factors, cytokines, and their cognate receptors are present at elevated levels in and around the fracture site [69].

Zinc has been shown to stimulate fracture healing. The role of zinc in fracture healing has been determined using the diaphyseal tissues obtained at 7 or 14 days after the fracture of femoral diaphysis of rats [68-71]. Protein content in the femoral-diaphyseal tissues was markedly increased in fracture healing, and many protein molecules were produced in the bone tissues with healing [70]. When the diaphyseal tissues with fracture healing were cultured, the significant increase in bone alkaline phosphatase activity and DNA content is caused. These increases were significantly enhaced after culture with zinc compound [70]. These increases were completely prevented after culture with an inhibitor of protein synthesis.

Culture with diaphyseal tissues of fracture healing caused a significant increase in IGF-I or TGF-β1 in culture medium [71]. The production of IGF-I or TGF-β1 from the bone tissues with fracture healing was significantly enhanced after culture with zinc compound. The effect of IGF-I or TGF-β1 in increasing protein content in the bone tissues with fracture healing was enhanced after culture with zinc compound [71].

When the femoral-diaphyseal tissues obtained from rats with fracture healing were cultured for 24 h in a serum-free medium, many proteins in the bone tissues were released into the culture medium [72]. Especially, a protein molecule of approximately 66 kDa was markedly increased with fracture healing. The increase in 66 kDa molecule was enhanced after culture with zinc compound [72]. The effect of zinc in increasing the 66 kDa molecule was based on a newly synthesized protein. Production of bone osteocalcin, which is significantly increased during fracture healing, has been shown to enhance after culture with zinc compound [73].

The results of N-terminal sequencing of 66 kDa protein showed that the N-terminus is identical to that of rat albumin [74]. The expression of albumin was seen in the diaphyseal (cortical bone) and metaphyseal (trabecular bone) tissues of rat femur. Albumin production in the bone tissues with fracture healing has been shown to increase after culture with PTH, IGF-I, or zinc compound [74]. Albumin has been demonstrated to express in bone marrow cells and osteoblastic MC3T3-E1 cells [75]. Albumin has been shown to stimulate the proliferation of osteoblastic MC3T3-E1 cells, and it suppresses alkaline phosphatase activity that is a marker enzyme of the differentiation of the cells *in vitro* [76]. Albumin has also been shown to have a suppressive effect on Runx2 (type 1) mRNA expression and a stimulatory effect on α1 (I) collagen mRNA level in osteoblastic MC3T3-E1 cells *in vitro* [77]. Albumin may play a physiological role in bone formation and mineralization in osteoblastic cells.

Thus, fracture healing induces a remarkable production of albumin, which has an anabolic effect on bone, in the femoral-diaphyseal tissues of rats. Zinc promotes fracture healing due to increasing many bone protein components including albumin, IGF-I, and TGF-β1 which can stimulate osteoblastic bone formation.

Bone fracture often occurs in osteoporosis. Chemical factors that can stimulate the healing of bone fracture have not been fully developed. The daily oral administration of zinc acexamate (100 mg Zn/kg) for 28 days caused a significant increase in calcium content, alkaline and acid phosphatase activities, protein and DNA contents in the femoral-diaphyseal tissues of rats with fracture healing [78]. The supplemental intake of zinc compound has a usefullness in the promoting of fracture healing of the femoral-diaphyseal tissues in rats.

ZINC COMPOUND PREVENTS BONE LOSS

β-Alanyl-L-histidinato zinc (II) (AHZ), in which zinc is chelated to β-alanyl-L-histidine, is a new zinc compound. Its molecular weight is 289.61. The oral administration of AHZ has been found to stimulate bone growth in weanling rats, and the compound has a potent effect in comparison with zinc sulfate [79]. AHZ has been demonstrated to stimulate bone formation in tissue culture [80] and proliferation and differentiation of osteoblastic cells *in vitro* [39, 40]. These effects of AHZ were potentially in comparison with that of zinc sulfate.

AHZ is easily absorbed from the intestine. The zinc in AHZ may largely accumulation in bone cells without difficulty because the metal binds to the hydroxyapatite of bone tissue [81]. AHZ can easily enter bone cells (osteoblasts and osteoclasts) in comparison with zinc sulfate [82]. This explains the different effects of AHZ and zinc sulfate on the cell functions.

AHZ has a potent stimulatory effect on protein synthesis in osteoblastic cells [83]. Aminoacyl-tRNA synthetase is an enzyme that synthesizes aminoacyl-tRNA, and the enzyme in osteoblastic cells was directly activated by AHZ (10^{-7} to 10^{-5} M). The effect of AHZ in increasing aminoacyl-tRNA synthetase activity was greater than that of zinc sulfate [84]. AHZ may be able to bind easily to aminoacyl-tRNA synthetase. β-Alanyl-L-histidine does not have an effect. Zinc ion is required in the revelation of AHZ action, since the action is completely abolished in the presence of dipicolinate, a chelator of zinc ion.

The effect of zinc sulfate, AHZ, di(N-acetyl-β-alanyl-L-histidinato) zinc, and di(histidino) zinc on bone metabolism is also compared *in vivo* and *in vitro* [83, 84]. The chemical form of zinc-chelating β-alanyl-L-histidine has been demonstrated to reveal a potent-anabolic effect on bone formation and calcification in some dipeptides used as ligand.

AHZ has been shown to have a preventive effect on bone loss with various pathophysiological conditions.

Zinc content in the cellular components, but not the matrix, has been demonstrated to be lower in the femoral diaphysis of aged rats (30 weeks old) than in that of weanling rats (3 weeks old) [85, 86]. Bone protein synthesis most likely deteriorates with increasing age. This deterioration has been found to restore after the oral administration of zinc (5 to 20 mg Zn/kg) for 3 days. The depletion of zinc in bone cells may cause the deterioration of bone formation in aged rats. The decrease in bone zinc with aging may play a role in the development of bone loss with increasing age. The supplementation of zinc may be important in the prevention of bone loss with aging. AHZ (10 – 75 mg/kg body weight/day) was orally administered to aged rats (30 weeks old). The administration caused a significant increase in zinc, calcium, and DNA contents in the femoral diaphysis [87]. The anabolic effect of AHZ has also been shown in the tissue culture using bone obtained from aged rats *in vitro* [88].

It is known that skeletal unloading caused by immobilization, spaceflight, bedrest, or hindlimb suspension results in osteopenia. Zinc content has been shown to be decreased in the femoral-metaphyseal tissues of rats with skeletal unloading [89, 90]. Skeletal unloading results in an inhibition of bone formation and induces an increae in bone resorption, and thereby decreasing in bone mass [91]. Skeletal unloading was designed using the model of hindlimb suspension in rats that we developed. Animals were fed for 4 days with the unloading. The unloading induced a significant increase in metaphyseal zinc content. Zinc accumulation into the metaphyseal tissues after a single oral administation of zinc sulfate (200 mg Zn/kg) was depressed after the unloading [89]. Serum zinc concentration in skeletal-unloaded rats was higher than that in normal rats. The impaired movement of zinc from serum into bone tissues may be caused with the unloading [89]. Skeletal unloading caused a significant decrease in bone components [89, 90]. Skeletal unloading-induced decrease in zinc content, alkaline phosphatase activity, and DNA content in the femoral diaphysis of rats was restored after the oral administration of AHZ (25-100 mg/kg/day) [92]. The effect of zinc has been shown to involve in newly synthesized proteins. The supplementation of zinc compound may have a role in the prevention of bone loss with skeletal unloading.

Aluminium has been shown to localize in the bone from patients with renal failure, and this induces bone disorders [93]. The dose of 2.7 and 5.4 mg Al/kg caused a significant increase in serum calcium concentration and bone acid phosphatase activity, while bone alkaline phosphatase activity and calcium content were not significantly altered [94]. The bone DNA content was significantly decreased with the dose of 5.4 mg Al/kg. These decreases of bone components were completely blocked after the simultaneous administration of AHZ (10 and 25 mg/kg). AHZ can prevent the revelation of the toxic effect of aluminium on bone metabolism in rats [94].

Feeding with low-calcium and vitamin D-deficient diets induces a decrease in serum calcium concentration and a corresponding fall in bone calcium content. When rats were fed on low-calcium (0.10%) and vitamin D-deficient diets for 14 days, the decrease in serum calcium and bone calcium was induced [95]. When AHZ (10-100 mg/kg/day) was orally administered to rats fed low-calcium and vitamin D-deficient diets for 14 days, the administration resulted in an increase in bone components [95]. The serum calcium and inorganic phosphorus concentrations were not significantly altered after AHZ administration. AHZ may directly stimulate bone formation independently of alteration in serum mineral homeostasis.

The inhibition of bone formation occurs during acute inflammation in the rat, and changes in osteoblast function are part of the acute phase response following local inflammation. Bone loss was seen in rats with adjuvant arthritis; local subcutaneous injection of 1% *Mycobacterium butyricum* suspension was used to induce systemic inflammatory response in rats. The decrease in calcium content and alkaline phosphatase activity were seen in the femoral diaphysis of rats with adjuvant arthritis [96]. The oral administration of AHZ (30 and 100 mg/kg) produced a significant increase in alkaline phosphatase activity, DNA, calcium and zinc contents in the femoral tissues of rats with adjuvant arthritis. AHZ did not have a direct inhibitory effect on inflammation. AHZ (300 mg/day) treatment has been shown to improve periarticular osteoporosis, probably through an increase of bone formation, in postmenopausal women with rheumatoid arthritis [97].

Glucocorticoid therapy induces the development of secondary hyperparathyroidism and the inhibition of osteoblastic function [98]. The steroid treatment caused a significant increase in serum PTH level, while serum calcium, inorganic phosphorus and zinc concentrations were not significantly altered [99]. The femoral-diaphyseal alkaline phosphatase activity, DNA and calcium contents were significantly decreased after the administration of steroid, although bone zinc content was not significantly changed. When AHZ (30 or 100 mg/kg) was orally administered for 30 days to rats administered with the steroid, the administration completely prevented the increase in serum PTH level and the decrease in the bone components caused by the steroid treatment. Bone zinc content was significantly increased after the administration of AHZ. AHZ may have a therapeutic effect in the steroid-induced osteoporosis.

Ovariectomy causes a lack of estrogen. Estrogen deficiency induces osteoporosis in humans and in rats [100, 101]. Ovariectomy caused a significant decrease in alkaline phosphatase activity, DNA and calcium contents in the femoral diaphysis of rats [102-104]. These decreases were completely prevented after the oral administration of AHZ (10-100 mg/kg/day) for 6 weeks [102]. AHZ restored bone loss in ovariectomized rats. Moreover, rats were fed for 3 weeks after ovariectomy, and then the animals were orally administered AHZ (10-100 mg/kg/day) for 3 months [103, 104]. A remarkable decrease in estrogen concentration in rat serum was seen 3 months after ovariectomy. The prolonged oral administration of AHZ did not influence serum estrogen level after ovariectomy. The administration of AHZ completely prevented reduction of bone mass [103]. AHZ administration for 3 months could completely prevent reduction in mineral content in the trabecular and cortical bone tissues of ovariectomized rats [104]. AHZ has been shown to have a preventive effect on ovariectomy-induced bone loss.

Zinc acexamate has been shown to have a potent effect on bone formation as compared with that of zinc sulafate and AHZ *in vitro* [105]. Zinc acexamate may be a good tool in therapy of osteoporosis. Bone loss has been shown to induce with diabetes [106, 107]. Streptozotocin (STZ)-diabetic rats induce an impairment of insulin secretion. A single subcutaneous administration of STZ (60 mg/kg body weight) to rats caused a significant decrease in body weight and serum zinc concentration, the increase in serum glucose and triglyceride levels, and the reduction of alkaline phosphatase activity, calcium and DNA contents in the femoral-diaphyseal and –metaphyseal tissues of STZ-diabetic rats [108, 109]. Zinc acexamate (25 mg Zn/kg) was orally administered once daily for 14 or 21 days to rats received a single subcutaneous administration of STZ. Zinc acexamate had potent-preventive effects on the changes in body weight, serum findings, and bone loss in STZ-administered rats, indicating its restorative effect on insulin-dependent (Type I) diabetic conditions [107, 109]. Zinc acexamate had a potent-preventive effect on diabetic conditions as compared with zinc sulfate. Zinc acexamate may have preventive and therapeutic effects on diabetes-induced bone loss in rats.

Zinc Supplementation and Osteoporosis Prevention

Osteoporotic patients have been shown to have lower levels of skeletal zinc than healthy individuals [19]. The reduction of levels of biological markers of nutrition in postmenopausal osteoporosis may be related to zinc deficiency. In postmenopausal women, urinary zinc has been used as a marker of bone resorption. Plasma and urinary zinc concentrations in 30 women with postmenopausal osteoporosis and in 30 healthy postmenopausal women, who served as controls, have been measured [110]. Plasma zinc levels did not differ between groups, but urinary zinc excretion has been found to be significantly higher in the women with postmenopausal osteoporosis [110]. The elevation of urinary zinc elimination in osteoporosis may be dependent on bone resorption [110], because zinc is located richely in bone tissues. Measurement of urinary zinc may be a usefull biochemical method of observing the positive clinical effect following alendronate or calcitonin therapy in postmenopausal women [111].

The relationship between indices of zinc nutritive status and biochemical markers of bone turnover in older adult European subjects has been shown [112]. A total of 387 healthy adults, aged 55-87 years was used in this study. There was seen some, albeit inconsistent, evidence of a relationship between zinc nutritive status and bone turnover in the older adult participants of the ZENITH study [112].

The supplements of trace minerals with or without calcium in post-menopausal women have been shown to have beneficial effects on bone density [20]. Low zinc intakes and reduced blood zinc concentrations have been reported to be associated with osteoporosis in women [113]. To examine the independent association between dietary zinc and plasma zinc and the association of each with bone mineral density (BMD) and 4 years-bone loss in community-dwelling older men. Of the original Rancho Bernardo Study subjects, 396 men (age: 45-92 years) were used. The mean dietary zinc intake was 11.2 mg. Dietary zinc intake and plasma zinc each have a positive association with BMD in men [113].

Protein undernutrition is frequent in the elderly. It contributes to the development of osteoporosis, possibly via lower IGF-I. Dietary zinc can influence IGF-I production. In the elderly, zinc supplementation accelerated the serum IGF-I response to essential amino acids-whey protein supplements by 1 week and decreased a biochemical marker of bone resorption [114].

Changes in circulating biochemical markers of bone metabolism in aged individuals with the food intake supplemented with zinc has been examined [156]. Sixty-three volunteers (31 men and 32 women) were divided into four groups of 15 or 16 male volunteers and 16 or 16 female volunteers, and each group was sequentially food containing 0.8 mg or 3.6 mg zinc once a day for 4 or 8 weeks as follows. The dietary intake of zinc for 8 weeks in men or women caused a significant increase in serum bone-specific alkaline phosphatase activity and γ-carboxylated osteocalcin concentration and a significant decrease in serum bone TRACP activity and N-telopeptide of type I collagen, as compared with the control group without dietary zinc [115]. This study suggests that the supplementation with zinc has a stimulatory effect on bone formation and an inhibitory effect on bone resorption in aged individuals.

Intake of dietary zinc may have a benefit effect in the prevention of osteoporosis.

CONCLUSION

Nutritional zinc is abundant in bone and may act as a local regulator of bone cells. Zinc plays a physiologic role in the regulation of bone homeostasis. Zinc stimulates osteoblastic bone formation and mineralization, and it inhibits osteoclastic bone resorption, thereby increasing bone mass [116-118]. Zinc tranporter has been shown to locate in osteoblastic cells [119] and osteoclastic cells [120]. Zinc transporter may be important role in the uptake, intracellular sequestration or efflux of zinc. Intracellular zinc stimulates the synthesis of many proteins at translational process due to activating aminoacyl-tRNA synthetase and to stimulate gene expressions in osteoblastic cells. Zinc also increaes DNA synthesis in the cells. The molecular mechanism of zinc on nuclear function in osteoblastic cells remains to be elucidated. In addition, the regulatory mechanism of intracellular zinc in osteoclastic cells has been poorly understood, although zinc stimulates apoptosis of osteoclastic cells which are mediated through Ca^{2+} signaling.

There is growing evidence that zinc finger transcription factors as zinc-binding protein play an important role in differentiation of osteoblastic cells and osteoclastic cells. A novel zinc finger-containing transcription factor, called Osterix (Osx), has been found in osteoblastic cells [121]. Osx is specifically expressed in all developing bones. Osx is required for osteoblast differentiation and bone formation, and it acts downstream of Runx2/Cbfa1. Cas-interacting zinc finger protein (CIZ) has been found to be a novel type inhibitor of bone morphogenetic protein (BMP)/Smad signaling in the modulation of BMP2-induced osteoblastic cell differentiation [122]. Moreover, Schnurri-3 (Shn3), a mammalian homolog of the Drosophila zinc finger adapter protein Shn, is an essential regulator of adult bone formation [123]. Runx2-mediated extracellular matrix mineralization is antagonized, revealing an essential role for Shn3 as a central regulator of postnatal bone mass. A novel TIZ (TRAF6-inhibitory zinc finger protein) has been shown to inhibit osteoclastogenesis and the function of tumor necrosis factor receptor-associated factor 6 [124]. TIZ may play a regulatory role during osteoclast differentiation by modulating TRAF6 signaling activity. Whether nutritional zinc state may affect on function of zinc finger proteins has been poorly understood. Presumably, zinc deficiency attenuates function of these proteins.

The decrease in bone zinc with pathophysiologic state has been shown to lead bone loss. The supplemental intake of zinc prevents bone loss. Zinc yeast has been shown to have a high bioavailability in rats and its dietary intake has an anabolic effect on bone components *in vivo* [125], indicating its usefulness as a functional food ingredient.

β-Alanyl-L-histidinato zinc (AHZ) or zinc acexamate has been shown to have a potent-anabolic effect in comparison with zinc sulfate. These compounds may be a clinical usefullness in the therapy of osteoporosis.

Thus, zinc supplementation may play an important role in the prevention and therapy of osteoporosis. Further development is expected.

References

[1] Prasad AS, Halsted JA, Nadimi M (1961) Syndrome of iron deficiency anemia, hepatosplenomegaly, hupogonadism, dwarfism and geophagia. *Am. J. Med.* 31:532-546

[2] Burch RE, Hahn HK, Sullivan JF (1975) Newer aspects of the roles of zinc, manganese, and copper in human nutrition. *Clin. Chem.* 21:501-520

[3] Parisi AF, Vallee BL (1969) Zinc metalloenzyme: Characteristic and significance in biology and medicine. *Am. J. Clin. Nutr.* 22:1222-1230

[4] Hsieh HS, Navia JM (1980) Zinc deficiency and bone formation in guinea pig alveolar implants. *J. Nutr.* 110:1581-1588

[5] Oner G, Bhaumick B, Bala RM (1984) Effect of zinc deficiency on serum somatomedin levels and skeletal growth in young rats. *Endocrinology.* 114:1860-1863

[6] Hurley LS (1981) Tetratogenic aspects of manganese, zinc, and copper nutrition. *Physiological Reviews.* 61:249-295

[7] da Cunha Ferreira RM, Marguiegui IM, Elizaga IV (1989) Tetratogenicity of zinc deficiency in the rat: Study of the fetal skeleton. *Tetratology.* 39:181-194

[8] Leek JC, Vogler JB, Gershwin ME, Golub MS, Hurley LS, Hendrickx AG (1984) Studies of marginal zinc deprivation in rhesus monkeys. V. Fetal and infant skeletal defects. *Am. J. Clin. Nutr.* 40:1203-1212

[9] Ronaghy HA, Reinhold JG, Mahloudji M, Ghavami P, Spirey Fox MR, Halstead JA (1984) Zinc supplementation of malnourished schoolboys in Iran: Increased growth and other effects. *Am. J. Clin. Nutr.* 40:1203-1212

[10] Herzberg M, Foldes J, Steinberg R, Menczel J (1990) Zinc excretion in osteoporotic women. *J. Bone Mineral Res.* 5:251-257

[11] Haumont S (1961) Distribution of zinc in bone tissues. *J. Histochem. Cytochem.* 9:141

[12] Hurley LS, Shyy-Hwa T (1972) Alleviation of tetratogenic effects of zinc deficiency by simultaneous lack of calcium. *Am. J. Phys.* 222:322-325

[13] Masters DG, Keen CL, Lonnerdal B, Hurley LS (1986) Release of zinc from maternal tissues during zinc deficiency or simultaneous zinc and calcium deficiency in the pregnant rat. *J. Nutr.* 116:2148-2154

[14] Murray EJ, Messer HH (1981) Turnover of bone zinc during normal and accelerated bone loss in rats. *J. Nutr.* 111:1641-1647

[15] Sherman SS, Smith JC, Tobin JD, Soares JH (1989) Ovariectomy, dietary zinc, and bone metabolism in retired breeder rats. *Am. J. Clin. Nutr.* 49:1184-1191
[16] Aitken JM (1976) Factors affecting the distribution of zinc in the human skeleton. *Calcif. Tissue Res.* 20:23-30
[17] Sauer GR, Wuthier RE (1990) Distribution of zinc in the avian growth plate. *J. Bone Mineral Res.* 5 (Suppl. 2):S162
[18] Lappalainer R, Knuuttila M, Lammi S, Alhava EM (1983) Fluoride content related to the elemental composition, mineral density and strength of bone in healthy and chronically diseased persons. *J. Chronic Disease.* 36:707-713
[19] Reginster JY, Strause LG, Saltman O, Franchimont P (1988) Trace elements and postmenopausal osteoporosis: A preliminary study of decreased serum manganese. *Med. Sci. Res.* 16:337-338
[20] Saltman P, Strause L (1991) Trace elements in bone metabolism. *J. Inorg. Biochem.* 43:284 (abstract)
[21] Parfitt AM (1990) Bone-forming cells in clinical conditions. In: Hall BK, editor. Bone Volume 1. *The Osteoblast and Osteocyte.* Boca Raton, FL: Telford Press and CRC Press, p.351-429
[22] Baron R, Vignery A, Horowitz M (1984) Lymphocytes, macrophages and the regulation of bone remodeling. *Bone Miner Res.* 2:175-243
[23] Canalis E, McCarthy T, Centrella M (1988) Growth factors and the regulation of bone remodeling. *J. Clin. Invest.* 81:277-281
[24] Yamaguchi M, Takahashi K (1984) Role of zinc as an acivator of bone metabolism in weanling rats. *Jpn J. Bone Metab.* 2:186-191
[25] Yamaguchi M, Inamoto K, Suketa Y (1986) Effect of essential trace metals on bone metabolism in weanling rats: Comparison with zinc and other metals` actions. *Res. Exp. Med.* 186:337-342
[26] Yamaguchi M, Uchiyama M (1987) Preventive effect of zinc for toxic actions of germanium and selenium on bone metabolism in weanling rats. *Res. Exp. Med.* 187:395-400
[27] Lai YL, Yamaguchi M (2005) Effects of copper on bone component in the femoral tissues of rats: Anabolic effect of zinc is weakened by copper. *Biol. Pharm. Bull.* 28:2296-2301
[28] Yamaguchi M, Osishi H, Suketa Y (1987) Stimulatory effect of zinc on bone formation in tissue culture. *Biochem. Pharmacol.* 36:4007-4012
[29] Yamaguchi M, Oishi H, Suketa Y (1988) Zinc stimulation of bone protein synthesis in tissue culture. Activation of aminoacyl-tRNA synthetase. *Biochem. Pharmacol.* 37:4075-4080

[30] Yamaguchi M, Matsui R (1989) Effect of dipicolinate, a chelator of zinc, on bone protein synthesis in tissue culture. The essential role of zinc. *Biochem. Pharmacol.* 38:4485-4489

[31] Shimokawa N, Yamaguchi M (1992) Characterization of bone protein components with polyacrylamide gel electrophoresis: Effects of zinc and hormones in tissue culture. *Mol. Cell. Biochem.* 117:153-158

[32] Yamaguchi M, Sakashita T (1986) Enhancement of vitamin D_3 effect on bone metabolism in weanling rats orally administered zinc sulphate. *Acta Endocrinol.* 111:285-288

[33] Yamaguchi M, Yamaguchi R (1986) Action of zinc on bone metabolism in rats. Increase in alkaline phosphatase activity and DNA content. *Biochem. Pharmacol.* 35:773-777

[34] Yamaguchi M, Inamoto K (1986) Differential effects of calcium-regulating hormones on bone metabolismin weanling rats orally administered zinc sulfate. *Metabolism.* 35:1044-1047

[35] McDonnell DP, Mongelsdorf DJ, Pike JW, Haussler MR, O'Malley BW (1987) Molecular cloning of complementary DNA encoding the avian receptor for vitamin D. *Science.* 235:1214-1217

[36] Yamaguchi M, Oishi H (1989) Effect of 1,25-dihydroxyvitamin D_3 on bone metabolism in tissue culture. Enhancement of this steroid effect by zinc. *Biochem. Pharmacol. 38*:3453-3459

[37] Yamaguchi M, Kitajima T (1991) Effect of estrogen on bone metabolism in tissue culture: Enhancement of the steroid effect by zinc. *Res. Exp. Med.* 191:145-154

[38] Lutz W, Burritt MF, Nixon DE, Kao PC, Kumar R (2000) Zinc increases the activity of vitamin D-dependent promoters in osteoblasts. Biochem. *Biophys. Res. Commun.* 271: 1-7

[39] Hashizume M, Yamaguchi M (1993) Stimulatory effect of β-alanyl-L-histidinato zinc on cell proliferation is dependent on protein synthesis in osteoblastic MC3T3-E1 cells. *Mol. Cell. Biochem.* 122:59-64

[40] Hashizume M, Yamaguchi M (1994) Effect of β-alanyl-L-histidinato zinc on differentiation of osteoblastic MC3T3-E1 cells: Increases in alkaline phosphatase activity and protein concentration. *Mol. Cell. Biochem.* 131:19-24

[41] Yamaguchi M, Hashizume M (1994) Effect of β-alanyl-L-histidinato zinc on proein components in osteoblastic MC3T3-E1 cells: Increases in osteocalcin, insulin-like growth factor-I and transforming growth factor-β. *Mol. Cell. Biochem.* 136:163-169

[42] Matsui T, Yamaguchi M (1995) Zinc modulation of insulin-like growth factor's effect in osteoblastic MC3T3-E1 cells. *Peptides.* 16:1063-1068

[43] Yamaguchi M, Fukagawa M (2005) Role of zinc in regulation of protein tyrosine phosphatase activity in osteoblastic MC3T3-E1 cells: Zinc modulation of insulin-like growth factor-I's effect. *Calcif Tissue Int.* 76:32-38

[44] Yamaguchi M, Kishi S, Hashizume M (1994) Effect of zinc-chelating dipeptides on osteoblastic MC3T3-E1 cells: Activation of aminoacyl-tRNA synthetase. *Peptides.* 15:1367-1371

[45] Yamaguchi M, Matsui T (1996) Stimulatory effect of zinc-chelating dipeptide on deoxyribonucleic acid synthesis in osteoblastic MC3T3-E1 cells. *Peptide.* 17:1207-1211

[46] Yamaguchi M, Goto M, Uchiyama S, Nakagawa T (2008) Effect of zinc on gene expression in osteoblastic MC3T3-E1 cells: enhancement of Runx2, OPG, and regucalcin mRNA expressions. *Mol. Cell. Biochem.* 312:157-66

[47] Yamaguchi M, Segawa Y, Shimokawa N, Tsuzuike N, Tagashira E (1992) Inhibitory effect of β-alanyl-L-histidinato zinc on bone resorption in tissue culture. *Pharmacology.* 45:292-300

[48] Yamaguchi M, Hashizume M (1994) Effect of parathyroid hormone and interleukin-1α in osteoblastic MC3T3-E1 cells: Interaction with β-alanyl-L-histidinato zinc. *Peptides.* 15:633-636

[49] Kishi S, Yamaguchi M (1994) Inhibitory effect of zinc compounds on osteoclast-like cell formation in mouse marrow cultures. *Biochem. Pharmacol.* 48:1225-1230

[50] Yamaguchi M, Kishi S (1996) Zinc compounds inhibit osteoclast-like cell formation at the earlier stage of rat marrow culture but not osteoclast function. *Mol. Cell. Biochem.* 158:171-177

[51] Yamaguchi M, Kishi S (1995) Inhibitory effect of zinc-chelating dipeptide on parathyroid hormone-stimulated osteoclast-like cell formation in mouse marrow cultures: Involvement of calcium signaling. *Peptides.* 16:629-633

[52] Zaidi M, Blair HC, Moonga BS, Abe E, Huang CL-H (2003) Osteoclastogenesis, bone resorption, and osteoclast-based therapeutics. *J. Bone Miner. Res.* 18:599-609

[53] Asagiri M, Takayanagi H (2007) The molecular understanding of osteoclast differentiation. *Bone.* 40:251-264

[54] Anderson DM, Marashovsky E, Billingoley WL, Dougall WC, Tometsko ME, Roux ER, Teepe MC, DuBose RF, Cosman D, Galibert L

(1997) A homologue of the TNF receptor and its ligand enhances T-cell growth an dentritic-cell function. *Nature.* (London) 390:175-179

[55] Yasuda H, Shima N, Nakagawa N, Yamaguchi K, Kinosaki M, Mochizuki S, Tomoyasu A, Yano K, Goto M, Murakami A, Tsuda E, Morinaga T, Higashio K, Udagawa N, Takahashi N, Suda T (1998) Osteoclast differentiation factor is a ligand for osteoprotegerin/osteoclastogenesis-inhibitory factor and is identical to TRANCE/RANKL. *Proc. Natl. Acad. Sci. U.S.A.* 95:3597-3602

[56] Tsuda E, Goto M, Mochizuki S, Yano K, Kobayashi F, Morinaga T, Higashio K (1997) Isolation of a novel cytokine from human fibroblasts that specifically inhibits osteoclastogenesis. *Biochem. Biophys. Res. Commun.* 234:137-142

[57] Inoue J, Ishida T, Tsukamoto N, Kobayashi N, Naito A, Azuma S, Yamamoto T (2000) Tumor necrosis factor receptor-associated factor (TRAF) family: adaptor proteins that mediate cytokine signaling. *Exp. Cell Res.* 254:14-24

[58] Yamaguchi M, Uchiyma S (2004) Receptor activator of Nf-κB ligand (RANKL)-stimulated osteoclastogenesis in mouse marrow culture is suppressed by zinc in vitro. *Int. J. Mol. Med.* 14:81-85

[59] Zou W, Hikim I, Tschoep K, Endres S, Zvi B-S (2001) Tumor necrosis factor-mediated RANK ligand stimulation of osteoclast differentiation by an autocrine mechanism. *J. Cell. Biochem.* 83:70-83

[60] Ma ZJ, Yamaguchi M (2000) Alteration in bone components with increasing age of newborn rats: role of zinc in bone growth. *J. Bone Miner. Metab.* 18:264-270

[61] Ma ZJ, Yamaguchi M (2001) Role of endogenous zinc in the enhancement of bone protein synthesis associated with bone growth of newborn rats. *J. Bone Miner. Metab.* 19:38-44

[62] Ma ZJ, Misawa H, Yamaguchi M (2001) Stimulatory effect of zinc on insulin-like growth factor-I and transforming growth factor-β1 production with bone growth of newborn rats. *Int. J. Mol. Med.* 8:623-628

[63] Gabbitas B, Pash J, Canalis E (1994) Regulation of insulin-like groth factor-II synthesis in bone cell cultures by skeletal growth factors. *Endocrinology.* 135:284-289

[64] Asahina I, Sampath TK, Nishimura I, Hauschka PV (1999) Human osteogenic protein-1 induces both chondroblastic and osteoblastic differentiation of osteoprogenitor cells derived from newborn rat calvaria. *J. Cell Biol.* 123:921-933

[65] Ma ZJ, Yamaguchi M (2001) Stimulatory effect of zinc and growth factor on bone protein component in newborn rats: Enhancement with zinc and insulin-like growth factor-I. *Int. J. Mol. Med.* 7:73-78
[66] Ma ZJ, Yamaguchi M (2001) Stimulatory effect of zinc on deoxyribonucleic acid synthesis in bone growth of newborn rats: Enhancement with zinc and insulin-like growth factor-I. *Calcif. Tissue.* Int. 69:158-163
[67] Barnes GL, Kostenuik PJ, Gerstenfeld LC, Eihorn TA (1999) Growth factor regulation of fracture repair. *J. Bone Miner. Res.* 14:1805-1815
[68] Eihorn TA (1998) The cell and molecular biology of fracture healing. *Clin. Orthop.* 355S:S7-S21
[69] Bolander ME (1998) Regulation of fracture repair by growth factors. *Proc. Soc. Exp. Biol. Med.* 200:165-170
[70] Igarashi A, Yamaguchi M (1999) Increase in bone protein components with healing rat fractures: Enhancement by zinc treatment. *Int. J. Mol.* Med. 4:615-620
[71] Igarashi A, Yamaguchi M (2001) Increase in bone growth factors with healing rat fractures: The enhancing effect of zinc. *Int. J. Mol. Med.* 8:433-438
[72] Igarashi A, Yamaguchi M (2002) Characterization of the increase in bone 66 kDa protein component with healing rat fractures: Stimulatory effect of zinc. *Int. J. Mol. Med.* 9:503-508
[73] Igarashi A, Yamaguchi M (2003) Great increase in bone 66 kDa protein and osteocalcin at later stage with healing rat fractures: Effect of zinc treament. *Int. J. Mol. Med.* 11:223-228
[74] Yamaguchi M, Igarashi A, Misawa H, Tsurusaki Y (2003) Enhancement of albumin expression in bone tissues with healing rat fractures. *J. Cell. Biochem.* 89:356-363
[75] Ishida K, Sawada N, Yamaguchi M (2004) Expression of albumin in bone tissues and osteoblastic cells: Involvement of hormone regulation. *Int. J. Mol. Med.* 14:891-895
[76] Ishida K, Yamaguchi M (2004) Role of albumin in ostoblastic cells: Enhancement of cell proliferation and suppression of alkaline phosphatase activity. *Int. J. Mol. Med.* 14:1077-1081
[77] Ishida K, Yamaguchi M (2005) Albumin regulates Runx2 and α1 (I) collagen mRNA expression in osteoblastic cells: Comparison with insulin-like growth factor-I. *Int. J. Mol. Med.* 16:689-694

[78] Igarashi A, Yamaguchi M (1999) Stimulatory effect of zinc acexamate administration on fracture healing of the femoral-diaphyseal tissues in rats. *Gen. Pharmacol.* 32:463-469

[79] Yamaguchi M, Ozaki K (1990) A new zinc compound, β-alanyl-L-histidinato zinc, stimulates bone growth in weanling rats. *Res. Exp. Med.* 190:105-110

[80] Yamaguchi M, Miwa H (1991) Stimulatory effect of beta-alanyl-L-histidinato zinc on bone formation in tissue culture. *Pharmacology.* 42:230-240

[81] Samuchson J (1967) Mechanism for the exchange of the calcium in bone mineral. *Nature.* 216:193-194

[82] Yamaguchi M, Ohtaki J (1991) Effect of beta-alanyl-L-histidinato zinc on osteoblastic MC3T3-E1 cells: Increases in alkaline phosphatase and proliferation. *Pharmacology.* 43:225-232

[83] Yamaguchi M, Kishi S (1994) Effect of zinc-chelating dipeptide on bone metabolism in weanling rats: Comparison with β-alanyl-L-histidinato zinc-related compounds. *Peptides.* 15:671-673

[84] Yamaguchi M, Kishi S (1994) Comparison of the effect of β-alanyl-L-histidinato zinc and its zinc-chelating ligand on bone metabolism in tissue culture. *Biol. Pharm. Bull.* 17:522-526

[85] Yamaguchi M, Ozaki K, Suketa Y (1989) Alteration in bone metabolism with increasing age: Effects of zinc and vitamin D_3 in aged rats. *J. Pharmacobio-Dyn.* 12:67-73

[86] Yamaguchi M, Ozaki K (1990) Aging affects cellular zinc and protein synthesis in the femoral diaphysis of rats. *Res. Exp. Med.* 190:295-300

[87] Yamaguchi M, Ozaki K (1990) Effect of the new zinc compound beta-alanyl-L-histidinato zinc on bone metabolism in elderly rats. *Pharmacology.* 41:345-349

[88] Kishi S, Yamaguchi M (1994) Stimulatory effect of β-alanyl-L-histidinato zinc on alkaline phosphatase activity in bone tissue from elderly rats: Comparison with zinc sulfate action. *Biol. Pharm. Bull.* 17:345-347

[89] Yamaguchi M, Ehara Y (1995) Zinc decrease and bone metabolism in the femoral-metaphyseal tissues of rats with skeletal unloading. *Calcif. Tissue Int.* 57:218-223

[90] Yamaguchi M, Ehara Y (1996) Effect of essential trace metal on bone metabolism in the femoral-metaphyseal tissues of rats with skeletal unloading: Comparison with zinc-chelating dipeptide. *Calcif. Tissue Int.* 59:27-32

[91] Morey ER, Baylink DJ (1978) Inhibition of bone formation during spaceflight. *Science.* 201:1138-1141

[92] Yamaguchi M, Ozaki K, Hoshi T (1990) β-Alanyl-L-histidinato zinc prevents skeletal unloading-induced disorder of bone metabolism in rats. *Res. Exp. Med.* 190:289-294

[93] Maloney NA, Ott SM, Alfrey AC, Miller NL, Coburn JW, Sherrard DJ (1982) Histological quantitation of aluminium in iliac from patients with renal failure. *J. Lab. Clin. Med.* 99:206-216

[94] Yamaguchi M, Ozaki K (1990) Beta-alanyl-L-histidinato zinc prevents the toxic effect of aluminium on bone metabolism in weanling rats. *Pharmacology.* 41:338-344

[95] Segawa Y, Tsuzuike N, Tagashira E, Yamaguchi M (1992) Preventive effect of β-alanyl-L-histidinato zinc on bone metabolism in rats fed on low-calcium and vitamin D-deficient diets. *Res. Exp. Med.* 192:213-219

[96] Segawa Y, Tsuzuike N, Itokazu Y, Tagashira E, Yamaguchi M (1993) Effect of β-alanyl-L-histidinato zinc on bone metabolism in rats with adjuvant arthritis. *Biol. Pharm. Bull.* 16:656-659

[97] Sugiyama T, Tanaka H, Kawai S (2000) Improvement of periarticular osteoporosis in postmenopausal women with rheumatoid arthritis by beta-alanyl-L-histidinato zinc: a pilot study. *J. Bone Miner. Metab.* 18: 335-338

[98] Pock WA (1984) The effects of glucocorticoids on bone cell metabolism and function. *Adv. Exp. Med. Biol.* 171:111-119

[99] Segawa Y, Tsuzuike N, Itokazu Y, Tagashira E, Yamaguchi M (1992) β-Alanyl-L-histidinato zinc prevents hydrocortisone-induced disorder of bone metabolism in rats. *Res. Exp. Med.* 192:317-322

[100] Johnston CC Jr, Hui SL, Witt RM, Appledonn R, Baker RS, Longcope C (1985) Early menopausal changes in bone mass and sex steroids. *J. Clin. Endocrinol. Metab.* 61:905-911

[101] Goulding A, Gold E (1987) Effects of chronic prednisolone treatment on bone resorption and bone composition in intact and ovariectomized rats receiving β-estradiol. *Endocrinology.* 122:482-487

[102] Segawa Y, Tsuzuike N, Tagashira E, Yamaguchi M (1993) Preventive effect of β-alanyl-L-histidinato zinc on the deterioration of bone metabolism in ovariectomized rats. *Biol. Pharm. Bull.* 16:486-489

[103] Yamaguchi M, Kishi S (1993) Prolonged administration of β-alanyl-L-histidinato zinc prevents bone loss in ovariectomized rats. *Jap. J. Pharmacol.* 63:203-207

[104] Kishi S, Segawa Y, Yamaguchi M (1994) Histomorphological confirmation of the preventive effect of β-alanyl-L-histidinato zinc on bone loss in ovariectomized rats. *Biol. Pharm. Bull.* 17:862-865
[105] Yamaguchi M, Gao YH (1998) Potent effect of zinc acexamate on bone components in the femoral-metaphyseal tissues of elderly female rats. *Gen. Pharmacol.* 30:423-427
[106] Hui SL, Epstein S, Johanston CC (1985) A prospective study of bone mass in patients with type I diabetes. *J. Clin. Endocrinol. Metab.* 60:74-80
[107] Mcnair P (1988) Bone mineral metabolism in human type I diabetes mellitus. *Dan. Med. Bull.* 35:109-121
[108] Yamaguchi M, Uchiyama S (2003) Preventive effect of zinc acexamate administration in streptozotocin-diabetic rats: Restoration of bone loss. *Int. J. Mol. Med.* 12:755-761
[109] Uchiyama S, Yamaguchi M (2003) Alteration in serum and bone component findings induced in streptozotocin-diabetic rats is restored by zinc acexamate. *Int. J. Mol. Med.* 12:949-954
[110] Relea P, Revilla M, Ripoll E, Arribas I, Villa LF, Rico H (1995) Zinc, biochemical markers of nutrition, and type I osteoporosis. *Age Ageing.* 24: 303-307
[111] Gur A, Colpan L, Cevik R, Nas K, Jale Sarac A (2005) Comparison of zinc excretion and biochemical markers of bone remodelling in the assessment of the effects of alendronate and calcitonin on bone in postmenopausal osteoporosis. *Clin. Biochem.* 38: 66-72
[112] Hill T, Meunier N, Andriollo-Sanchez M, Ciarapica D, Hininger-Favier I, Poloto A, O'Connor JM, Coudray C, Cashman KD (2005) The relationship between the zinc nutritive status and biochemical markers of bone turnover in older European adults: the ZENITH study. *Eur. L. Clin. Nutr.* 59 (Suppl 2): S73-S78
[113] Hyun TH, Barrett-Connor E, Milne DB (2004) Zinc intakes and plasma concentrations in men with osteoporosis: the Rancho Bermardo Study. *Am. J. Clin. Nutr.* 80: 715-721
[114] Rodondi A, Ammann P, Ghilardi-Beuret S, Rizzoli R (2009) Zinc increases the effects of essential amino acids-whey protein supplements in frail elderly. *J. Nutr. Health Aging.* 13: 491-497
[115] Yamaguchi M, Igarashi A, Sakai M, Degawa H, Ozawa Y (2005) Prolonged intake of dietary fermented isoflavone-rich soybean reinforced with zinc affects circulating bone biochemical markers in aged individuals. *J. Health Sci.* 51: 191-196

[116] Yamaguchi M (1995) β-Alanyl-L-histidinato zinc: A potent activator in bone formation. *Curr. Med. Chem.* 1:356-365

[117] Yamaguchi M (1995) β-Alanyl-L-histidinato zinc and bone resorption. *Gen. Pharmac.* 26:1179-1183

[118] Yamaguchi M (1998) Role of zinc in bone formation and bone resorption. *J. Trace Elem. Exp. Med.* 11:119-135

[119] Inoue K, Matsuda K, Itoh M, Kawaguchi H, Tomoike H, Aoyagi T, Nagai R, Hori M, Nakamura Y, Tanaka T (2002) Osteopenia and male-specific sudden cardiac death in mice lacking a zinc transporter gene, Znt5. *Hum. Mol. Genet.* 15:1775-1784

[120] Khadeer MA, Sahu SN, Bai G, Abdulla S, Gupta A (2005) Expression of the zinc transporter ZIP1 in osteoclasts. *Bone.* 37:296-304

[121] Nakashima K, Zhou X, Kunkel G, Zhang Z, Deng JM, Behringer RR, de Crombrugghe B (2002) The novel zinc finger-containing transcription factor osterix is required for osteoblast differentiation and bone formation. *Cell.* 108:17-29

[122] Shen ZJ, Nakamoto T, Tsuji K, Nifuji A, Miyazono K, Komori T, Hirai H, Noda M (2002) Negative regulation of bone morphogenetic protein/Smad signaling by Cas-interacting zinc finger protein in osteoblasts. *J. Biol. Chem.* 277: 29840-29846

[123] Jones DC, Wein MN, Oukka M, Hofstaetter JG, Glimcher MJ, Glimcher LH (2006) Schnur Regulation of adult bone mass by the zinc finger adapter protein ri-3. *Science.* 312: 1223-1227

[124] Shin JN, Kim I, Lee JS, Koh GY, Lee ZH, Kim HH (2002) A novel zinc finger protein that inhibits osteoclastogenesis and the function of tumor necrosis factor receptor-associated factor 6. *J. Biol. Chem.* 277:8346-53

[125] Yamaguchi M, Igarashi A, Uchiyama S (2004) Bioavailability of zinc yeast in rats: Stimulatory effect on bone calcification in vivo. *J. Health Sci.* 50:75-81

INDEX

A

B

C

D

G

H

I

N

R

S

T

U

V

W

X

Y

Z